Springer Series in Optical Sciences   Volume 48

Edited by Theodor Tamir

W0037417

Springer Series in Optical Sciences

Editorial Board: J.M. Enoch D.L. MacAdam A.L. Schawlow K. Shimoda T. Tamir

Volumes 1 – 41 are listed on the back inside cover

Integrated Optics

Proceedings of the
Third European Conference, ECIO'85
Berlin, Germany, May 6–8, 1985

Editors
H.-P. Nolting and R. Ulrich

With 195 Figures

Springer-Verlag Berlin Heidelberg GmbH

Dr. Hans-Peter J. Nolting

Heinrich-Hertz-Institut für Nachrichtentechnik Berlin GmbH,
Einsteinufer 37, D-1000 Berlin 10, Germany

Professor Dr. Reinhard Ulrich

Technische Universität Hamburg-Harburg, Harburger Schloß-Straße 20,
D-2100 Hamburg 90, Fed. Rep. of Germany

Programme Committee
Reinhard Ulrich *(Chairman)* (TU Hamburg-Harburg, D)
Andreas Schlachetzki *(Co-Chairman)* (HHI Berlin, D)
Franz Auracher (Siemens AG, Munich, D)
Alain Carenco (CNET, Bagneux, F)
Paul Lagasse (University of Gent, B)
John E. Midwinter (Univ. College London, London, GB)
Dan B. Ostrowsky (Université de Nice, F)

Michel Papuchon (Thomson CSF, Orsay, F)
Klaus Petermann (TU Berlin, D)
Franz-K. Reinhart (EPFL Lausanne, CH)
Wolfgang Sohler (Universität Paderborn, D)
Stefano Sottini (IROE, Florenz, I)
William J. Stewart (Plessey, Towcester, GB)
Lars H. Thylén (Ericsson, Stockholm, S)
Chris D. W. Wilkinson, (Glasgow Univ., Scottland, GB)

Local Organization Committee
Andreas Schlachetzki, *(Chairman)* HHI Berlin
Hans-Peter Nolting, *(Co-Chairman)* HHI Berlin
Immanuel Broser, TU Berlin

Anton Heuberger, FhG Berlin
Klaus Petermann, TU Berlin
Reinhard Ulrich, TU Hamburg-Harburg

Secretary: Ingrid Weber-Zuckarelli (Heinrich-Hertz-Institut, Berlin, D)

Sponsors
DGaO Deutsche Gesellschaft für Angewandtee Optik
DPG   Deutsche Physikalische Gesellschaft
DPG   Regionalverband Berlin

DPG Arbeitsgemeinschaft Quantenoptik
IEE   The Institution of Electrical Engineers (IEE)
Electronic Division

Financial support was optained from: Senator für Wissenschaft und Forschung, Berlin

ISBN 978-3-662-13571-6   ISBN 978-3-540-39452-5 (eBook)
DOI 978-3-540-39452-5

2153/3130-543210

Preface

The development of miniaturized and ruggedized optical circuits, containing a number of optical and perhaps also electronic components integrated on the same substrate, and performing useful optical functions - this is the goal of the key technologies for future systems of communication, of instrumentation, and of general signal processing; it is expected to combine and to complement the established technologies of microelectronics, optoelectronics, and fiber-optics.

Today, after more than fifteen years of research on integrated optics, this goal appears to be almost within reach. The theoretical problems of light propagation and of numerous forms of coupling and interactions in integrated-optical structures are generally well understood. A great variety of single components for integrated optics has been demonstrated experimentally, and more recently also the successful integration of several components on a common substrate. Laboratory operation of such integrated-optical 'chips' has been reported, e.g., for RF spectrum analysis, for high-speed analog/digital conversion, for a fiber-optic gyro, and for various high-performance semiconductor laser sources.

Before commercial fabrication and technical application of such devices can take place, however, their performance has to be further improved. Serious technological and material problems are still to be overcome which are related to the small transverse dimensions and high optical power densities typical for integrated-optical waveguides. Progress can be expected here by further improvements and diversifications of micro-fabrication technologies and (perhaps more efficiently) by learning how to better adapt the optical structures to the existing technologies.

Considerable potential for further progress of integrated optics exists also in the field of nonlinear optics, notably for frequency conversion and for optical logics. In the past, this field had received relatively little attention, but more recently, a number of promising results have been obtained.

The three topics mentioned - technology, materials, and nonlinear optics - and in addition a variety of interesting components, these are the main subject areas to be discussed at the Third European Conference on Integrated Optics, to be held in Berlin during May 6-8, 1985. The present volume of Proceedings of this conference contains the manuscripts of contributed and invited papers to be presented at that conference. The sequence and grouping of these papers correspond to the programme of the conference, dictated heavily by the need to arrange reasonably self-contained sessions of prescribed lengths. All manuscripts were supplied camera-ready by the authors and are repoduced here without further proof. This procedure emphasizes speed rather than accuracy of publication; it has been adopted so as to permit a relatively late deadline and yet have the book available as the conference digest during the conference.

It is a pleasure to thank all those who contributed to this volume: All authors who prepared their manuscripts in the recommended style, all members of the ECIO programme committee for their advice and comments, Mr. R. Michels of Springer-Verlag for the unproblematic cooperation, numerous colleagues from the Heinrich-Hertz Institute for their expert assistance, and in particular Mrs. I. Weber-Zuckarelli, the Conference Secretary, who efficiently handled a remarkable amount of organizational work in compiling this book. Finally we wish to acknowledge the generous financial support by the Senator für Wissenschaft und Forschung of the State of Berlin which was very essential for the realization of this Conference.

Hamburg, Berlin
February 1985

R. Ulrich
H.-P. Nolting

Contents

*Invited paper

Part III   Semiconductor-Devices

Part IV   Modulators

Part V   Fundamentals and Waveguiding

Part I

Applications

Monomode Active Fibre Devices: The Effects of Overlay Index

B.K. Nayar

British Telecom Research Labs. Martlesham Heath, Ipswich
Suffolk, IP5 7RE, United Kingdom

The attenuation characteristics of a monomode half-coupler block with
change in the oil overlay refractive index have been measured. The
index change required for cut-off modulation (>20 dB) is 0.0008.

A number of monomode passive fibre devices, STOLEN [1], such as directional
couplers, isolators, polarisers, polarisation controllers etc have been
developed for use in both fibre sensors and communication systems. The
advantages of these devices compared to integrated-optic components are,
simplicity, ease of splicing, low insertion loss and high performance.
Over the past year there have been reports on the realisation of Raman
amplifiers, DESURVIRE [2], and at present other active components are being
considered.

In this paper we report on measurements carried out on polished half-
couplers using oil overlay to investigate the potential of active fibre
devices with electro-optic/non-linear material overlays for use as phase/
amplitude modulators and sensing.

The half-coupler block consist of a fused silica block with a curved
groove in which a monomode fibre is rigidly fixed using an epoxy. The fibre
cladding is then lapped to within a micron from the core, NAYAR [3], see
Fig 1. From simple theoretical considerations it is expected that:

(i) the effect of varying the overlay index to values less than the core
index will be to modulate the phase of the fibre mode. The phase change
will arise due to change in the effective mode index.

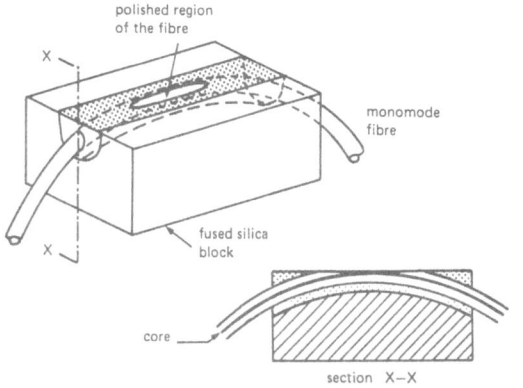

Fig. 1. Schematic of the half-coupler block

(ii) the effect of increasing the overlay index to above the effective mode index will result in the mode being radiated into the overlay. The rate of change of phase or attenuation with the index change will depend upon the effective mode index being more sensitive at low fibre V-values. Our measurements address the change of the overlay index required for these effects.

The experimental arrangement used to investigate the effect of the overlay index on transmission through the half-coupler block is shown in Fig 2. The cell for the oil was waxed onto the block surface. The refractive index of the oil was varied by cooling from a high temperature and was measured using a thermocouple. The measurements were carried out at 1.3 µm and 1.52 µm wavelengths with a block having a curved groove of 25 cm radius and at 1.52 µm with a block having a 75 cm radius groove.

In Fig 3 attenuation in decibels is plotted as a function of the change in temperature from some arbitrary temperature depending on the index of the oil used. The rate of change in attenuation with temperature for 25 cm and 75 cm blocks at 1.52 µm wavelengths are -15.1 dB/°C and -18.5 dB/°C

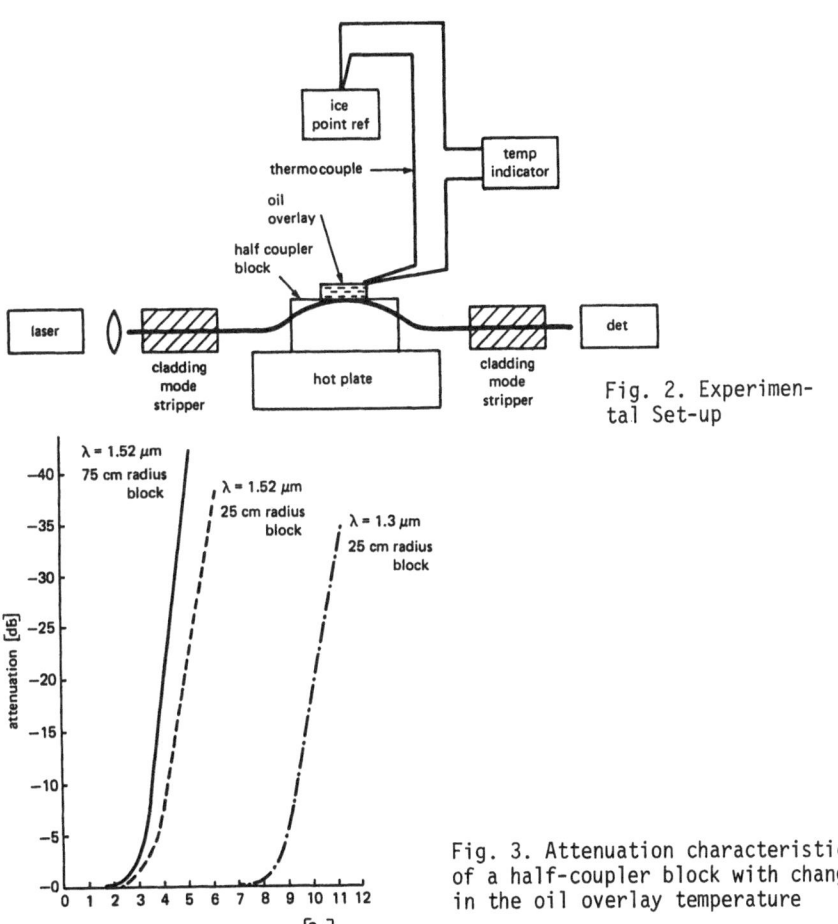

Fig. 2. Experimental Set-up

Fig. 3. Attenuation characteristics of a half-coupler block with change in the oil overlay temperature

3

respectively. The oil (Cargille Inc) has a rate of index change with temperature, dn/dT, of $-3.7 \times 10^{-4}/°C$ at the Sodium D-line. It can be seen from figure 3 that at 1.52 μm wavelength for the 25 cm radius block the attenuation goes up from under 0.1 dB to 20 dB for 3°C change in the temperature, while for the 75 cm radius block 2.1°C change in temperature is necessary for the same attenuation. For both these blocks, approximately 1 μm of the cladding was left unlapped. The effective interaction lengths in the two case are approximately 0.8 mm and 1.4 mm respectively. The other interesting feature of the characteristic is that for larger attenuation the relative increase in temperature required is very small. The index change, referred to the Sodium D-line, required for 20 dB attenuation is thus of the order of 11.1×10^{-4} for the 25 cm radius block and 7.8×10^{-4} for 75 cm block at 1.52 μm wavelength.

The attenuation characteristic of 1.3 μm wavelength is similar, except that the mode cut-off occurs for a much greater value of index change. This is to be expected as the effective mode index is much higher at 1.3 μm. The difference in the effective mode indices measured in this manner agrees well with the computed values.

In conclusion, these measurements show that active overlays could be used to realise monomode fibre phase/cut-off modulators, wavelength de-multiplexer and in sensors where one wavelength can be used for sensing while the other is used to determine the system status. The index change required is obtainable with existing electro-optic materials. At the meeting we will report on both the experimental results and theoretical behaviour and consider device designs. The author will like to acknowledge G.E. Clements for the coupler block fabrication, G. Harold for the assistance with measurements, D.R. Smith for comments, and thank the Director of Research, British Telecom, for permission to publish this work.

References

1. R.H. Stolen: "Single-mode Fiber Devices", paper ThC1, in Tech. Digest of Topical Mtg in Integrated and Guided Wave Optics Conf., IEEE (1984) and references therein.
2. E. Desurvire, K. Papuchon, J.P. Pocholle, J. Raffy and D.B. Ostrowsky: Electron. Letts., 19, 751 (1983).
3. B.K. Nayar and D.R. Smith: Optics Letters, 8, 543 (1983).

Vibration Sensor Using Integrated Optical Coupling Element

S. Honkanen, S. Tammela, P. Koivisto, and M. Leppihalme

Technical Research Centre, Semiconductor Laboratory, Otakaari 5 A
SF-02150 Espoo, Finland

M. Mäklin

Central Laboratory, Imatra Co., Ltd. P.O. Box 112, SF-01601 Vantaa, Finland

A vibration sensor based on a cantilever fiber and an integrated optical coupling element is presented. The main advantage of the device is its insensitivity to intensity variations of the source.

1 Introduction

A number of principles can be used in fiber-optic vibration sensors. The simplest approach is the coupling of light from a cantilever fiber to a fixed fiber. This approach has been used as a hydrophone [1] and as an accelerometer [2]. In this work experimental results on a new kind of fiber-optic vibration sensor utilizing integrated optical waveguides are given. The main idea is to measure the light coupled from a cantilever fiber to integrated optical waveguides. The use of waveguides gives several advantages. The shape of the light channels can be tailored to suit different kinds of fibers. Two separate waveguides can be placed near each other so that the power ratio can be measured instead of the power level as proposed in ref. [3]. Furthermore, with a proper design the sensor is sensitive only to one direction.

The paper covers the sensor structure, waveguide fabrication, measured power coupled into light channels as a function of the displacement of the fiber and the demonstration of the sensor as a vibrometer.

2 Coupler and measurement set-up

The structure of the coupling device and the measurement set-up are presented schematically in fig 1. The waveguides were fabricated into glass substrates using two-step silver-sodium ion exchange technique explained

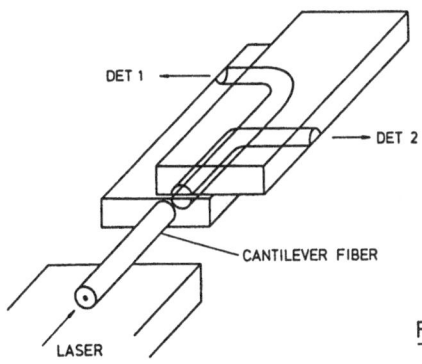

Fig.1 Coupler and measurement set-up

in detail in ref. [4]. Vacuum evaporated silver stripes were used as ion sources. After electric field assisted ion migration and a drive-in diffusion the waveguides were approximately 20 µm deep and 50 µm wide. To form an integrated coupling element two separate guides were glued together with silicon rubber [5], which effectively isolates the waveguides from each other.

The properties of the proposed sensor were measured using a single-mode fiber (core diameter 6,5 µm) fabricated in the laboratory. Light from a HeNe-laser was launched into the fiber through a 20x microscope objective. The output end of the fiber was adjusted very close to the common end of the waveguides. The separation was approximately 10 µm. Large step index fibers were butt-jointed to the output ends of the waveguides. The light from each channel was detected by Si photodiodes connected to digital voltmeters and a microcomputer.

3 Results

Since the operation of the device is based on change, in coupling of light between the fiber and the two waveguides, we first measured the coupling as a function of transverse displacement of the fiber. The graph in fig. 2 represents the measured intensities from the waveguides. The difference/sum of the intensities is shown in fig. 3. The effect of intensity variations of the light source was examined by repeating the measurement described above. During the measurement the coupling between the laser and the fiber was arbitrarily changed. Results are shown in figs. 4 and 5. It can be seen that the difference/sum of the detected intensities remains unaffected. Also, it was observed that the lateral displacement variations of the cantilever fiber were efficiently eliminated. The measured curves remained unchanged within lateral displacements of 10 µm.

These results confirm the benefits of using two collecting waveguides instead of one fiber. The error, which limits the sensitivity in practice, originating from the light source noise and fluctuations in coupling light to the single-mode fiber is considerably reduced. The construction is also

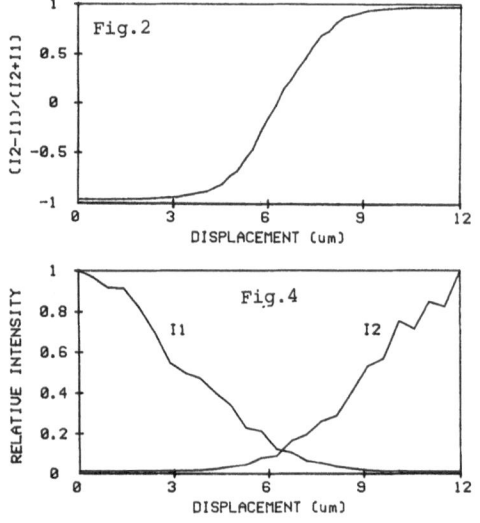

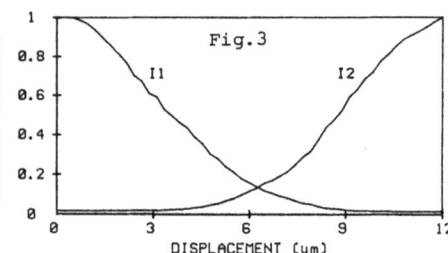

Fig. 2. Measured intensities from the two channels as a function of transverse displacement
Fig. 3. Difference/sum using the same data as in Fig. 2.
Fig. 4. The same as in Fig. 2. but the coupled power to the single-mode fiber was varied

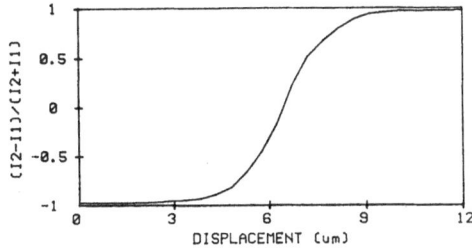

Fig. 5. Difference/sum using the same data as in Fig. 4.

simple, although the adjustment of the cantilever fiber is somewhat diffi-cult. The dynamic range and the sensitivity of the device can be further improved by decreasing the thickness of the glue between the waveguides, which in this case was approximately 5 μm.

In order to demonstrate the vibration sensor, the fiber was vibrated by touching the device. The outputs of the photodiodes were connected to an oscilloscope, and the signals shown in fig. 6 were obtained. The resonant frequency was about 60 Hz. Of course, in a proper sensor design the reso-nant frequency should be adjusted and damped [2].

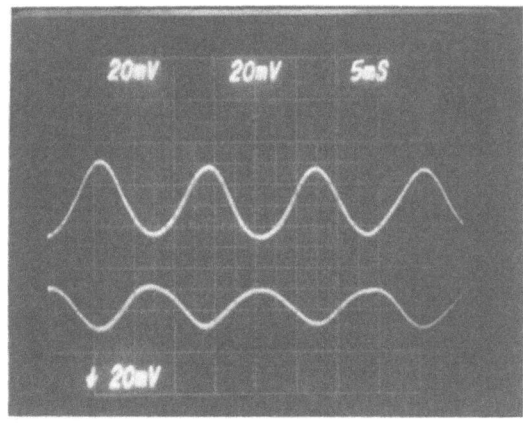

Fig. 6. The output signals when the cantilever fiber was vibrated

4 Conclusion

The use of integrated optical waveguides in connection with cantilever fiber has been demonstrated as a displacement and vibration sensor. The device is unique in that the light is coupled from the fiber to two sepa-rate waveguides so that the power ratio can be measured instead of the power level. Therefore the sensitivity to light source fluctuations is remarkably reduced. Before the physical limits of the sensitivity can be calculated more work has to be done concerning for example the modal noise of the waveguides and the output fibers.

References

1. W.B. Spillman, Jr. and R.L. Gravel: "Moving fiber-optic hydrophone", Opt.Lett. 5, 30(1980).

2. G.A. Rines: "Fiber-optic accelerometer with hydrophone applications",
   Appl.Opt. 20, 3453(1981).
3. R.A. Soref and D.H. McMahon: "Tilting-mirror fiber-optic accelerometer", Appl.Opt. 23, 486(1984).
4.   J. Viljanen and M. Leppihalme: "Fabrication of optical strip waveguides with nearly circular cross-section by silver ion migration
   technique", J.Appl.Phys. 51, 3563(1980)
5.   J. Kurki and J. Viljanen: "Integrated optics coupling element for
   backscatter measurements", Proc. 7th ECOC Copenhagen, Denmark
   (1981), post deadline paper 13.5.

Digital Optical Computing

A.W. Lohmann

Physikalisches Institut der Universität, Erwin-Rommel-Straße 1
D-8520 Erlangen, Fed. Rep. of Germany

1. Abstract

Photons do not interact, unless they meet in a suitable nonlinear material. Within that material it is possible to perform logical operations on two light beams or store information by means of optical bistability. Outside of such a material light beams are well suited to carry out the communication jobs of computing, especially for highly parallel computing.

The fundamentals of optical computing will be explained, a laboratory experiment will be described, and it will be discussed how integrated optics might contribute to digital optical processing.

2. Introduction

The development of the digital <u>electronic</u> computer was highly successful and its capabilities are still improving at a brisk rate. Are there any chances for digital <u>optical</u> computing? Indeed, there exist reasons to believe that an optical system can compete with electronic computing in some situations. We will analyze in the following first the basic aspects of optical computing and we will then describe as a laboratory example an optical, logical, parallel processor based on diffraction or scattering effects.

Finally, we will speculate how integrated optics might contribute to optical computing.

3. Aspects of Optical Computing

Certainly, photons can be used as carriers of information as well as electrons. However, the physical properties of both are quite different. The interaction between two electrons is strong, whereas two photons, normally, do not interact at all. Therefore, electron signals are easy to switch which is necessary to perform logic operations. But also for photons there exist now several approaches to perform logical operations /1,2/. A major breakthrough was, for example, the development of fast optical switches based on nonlinear materials /3/.

These new components can be adapted also to act as fast memory or as signal repeater. The point to remember here is: if two electrons are getting close

9

together, they will always interact. Two photons, on the other hand, will interact only if they meet within a special piece of exotic material. Hence, a photonic system can be designed such that interaction takes place only where it is wanted. Electrons, on the other hand, interact everywhere, even where interaction is undesirable.

Another aspect of computing systems is the communication between components. This aspect may be crucial for further progress in large-scale computing, since it becomes increasingly difficult to improve the speed of the components. Hence, progress will rely on computing with many processors in parallel. Parallel processing poses communications problems. Because of low interaction, photons are especially well suited for communication jobs. Electrons have to be guided carefully by means of wires and these wires have to be kept apart. But photons can travel in free space, even with intersections, without any guiding structure. This capability allows a new type of architecture and facilitates especially the design of parallel processing systems which require a very complex interconnection network /4/. Conventional, electronic computers are limited in this respect by the so-called "von Neumann" bottleneck. In these systems the number of interconnections is reduced at the price of a serial addressing scheme.

A specific communications network is the so-called "perfect shuffle", which can be implemented optically quite efficiently /7/.

As a last argument for an optical computing system we may recall that there is no direct influence of electromagnetic fields on photons. By use of optical components a computer could be hardened against electromagnetic fields, a property which is often desirable in an industrial environment.

From all these arguments we may conclude that in principle an optical, digital parallel computer is possible and desirable. Best opportunities for applications can be found in the field of parallel processing with high data rates (e.g., in a special purpose processor) or as system hardened against electromagnetic fields. Parallel processing will be profitable especially if the data set is parallel by its semantic nature. Image processing is an example.

4. Optical, logical operations in parallel

Now we want to mention one example for the implementation of parallel, logic operations by optical means. Binary, logical operations in parallel have been realized, e.g., by use of a LCLV /1/ or by shadow casting /2/. Our system uses encoded data which are then processed by simple, spatial filtering. The original, binary data with two logic levels are encoded on a spatial carrier, which may ·be deterministic (i.e., a grating structure) or stochastic (i.e., a diffuser structure) /5,6/. In the Fourier plane of an optical system the data of different logic levels are then separated. By proper choice of two spatial filters in two cascaded 4-f systems all 16 binary logical operations can be implemented. Even higher order logic with more than two logic levels can be realized. At the output the different logic levels are distinguished by intensity or by color. The unequal input/output data representation as well as the need for preprocessing is a disadvantage, but the logic operations themselves are performed with a high degree of parallelism.

5. Opportunities for integrated optics

The technology of integrated optics and the technology as used for optical bistability are similar in many ways. In addition there are some specific opportunities for using existing 10-components in an optical processor. For example an one-dimensional array of N/2 Delta-Beta couplers may perform the exchange-or-bypass operations as needed for a perfect shuffle network, that handles N channels in parallel. The shuffling itself could be done easily with free-space optics /7/. The Delta-Beta array would allow the exchange of positions between adjacent channels. Going 3logN times through such a network allows one to produce any wanted new permutation of data channels. The free-space optical part of this system works also with a two-dimensional array of channels, say $N^2 = (1024)^2$. A stack of N one-dimensional Delta-Beta arrays, each with N/2 components, would be needed to match the free-space part of the whole network. The purely-optical part can perform one cycle on N^2 data within one nanosecond. Hence, the permutation of a million channels can be achieved in less than a microsecond if the Delta-Beta couplers can be switched at nanosecond rates.

References

1. A.A. Sawchuk and T.C. Strand, Proc. IEEE 72 (1984) 758
2. Y. Ichioka and J. Tanida, Proc. IEEE 72 (1984) 787
3. H.M. Gibbs, S.L. McCall and T.N.C. Venkatesan, Opt. Engin. 19 (1980) 463
4. A. Huang, Proc. IEEE 72 (1984) 780
5. H. Bartelt, A.W. Lohmann and E.E. Sicre, J. Opt Soc. Am. A1 (1984) No. 9
6. A.W. Lohmann and J. Weigelt, Opt. Comm. 52 (1984) 255
7. A.W. Lohmann, W. Stork and G. Stucke, "Perfect shuffle optical network", to be published in Appl. Opt.

Step-Index Profile Microlenses Formed in Glass

B.G. Pantchev, A.G. Kebedjiev*, and I.T. Savatinova

Institute of Solid State Physics, *Institute of Electronics, Bulgarian Academy of Sciences, 1184 Sofia, Lenin Blvd. 72, Bulgaria

A step-index distribution in planar arrayed microlenses is obtained by field-assisted sodium-silver ion exchange in glass. The optical characteristics of the lenses depend only on the lens geometry and the value of the refractive index change.

1. Introduction

Mass-producibility of lightwave components is a substantial requirement for their application in the optical fiber communications and optoelectronic systems. However, the conventional microoptic and planar waveguide elements need a very precise alignment to fibers. To overcome this, IGA et al. /1/ have proposed the concept of stacked planar optics including 2-D arrays of microlenses and other optical devices. Using selective electromigration of Tl ions into glass substrates they have obtained distributed-index microlenses /2,3/. In order to avoid the use of toxic compound, such as Tl-salts, we have fabricated planar arrayed microlenses by a field-assisted sodium-silver ion exchange process /4/. In the present paper we discuss the lens two-dimensional index profiles in connection with the lens properties and the technological parameters.

2. Microlens Technology

Microlenses are fabricated in glasses by a field-enhanced Na-Ag ion exchange in 90 mol % $NaNO_3$ and 10 mol % $AgNO_3$ melts. We used 4.2 mm thick substrates of sodalime glasses ($n_D=1.515$) with composition: SiO_2 (72.58%), Na_2O (15.33%), CaO (6.71%), MgO (3.87%), Al_2O_3 (1.25%), Fe_2O_3 (0.17%) and K_2O (0.08%). A 10 x 10 array of circular windows is defined in the deposited thick Ti-layer by means of conventional photolithographic techniques. The window diameter D_w varied from 0.04 to 0.5 mm. A constant electric field of 12-24 V/cm during 4-1.5 hours at 621 K was used.

The formation of the lenses is based on the fact that during the exchange process through the circular windows opened in the mask layer, a "side diffusion" of the penetrating ions occurs.

The resulting lenses have diameters D_1 of 0.4-1.6 mm and focal lengths f=2-8 mm (in air). Typical values of the numerical aperture NA are 0.1-0.17 and the focal spot diameter 4-15 μm.

3. Refractive Index Profiles

Refractive index profiles are obtained by reflected power measurments /5,6/ using the experimental setup shown in Fig. 1. The beam of a He-Ne laser L (6328 $\overset{o}{A}$) is focused by a microscope objective 40 x/0.65 (MO) onto the polished surface of the sample. The surface is adjusted normally to the incident beam with an accuracy of 0.14^o and within $\pm$ 2 μm from the focal plane. The spatial resolution is 4 μm. To cope with laser stability problems a dual channel photometer (DCPM) was employed measuring the log ratio of reflected and reference optical powers.

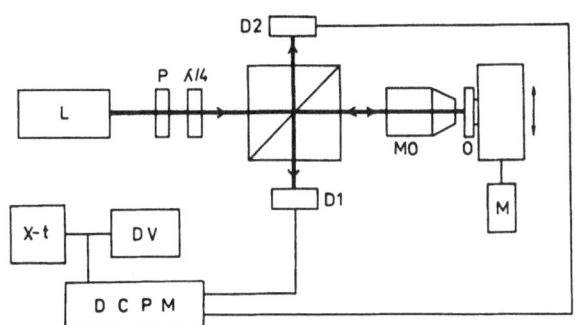

Fig.1

Two-dimensional profiles were investigated. Fig. 2 shows the index profile measured at the surface of a lens with the following parameters: D_1=660 μm, f=2.2 mm and NA=0.15. This profile is directly related to the lateral index distribution. When scanning a longitudinally cut lens along the x-direction, the depth index distribution is obtained (Fig. 3). Both profiles have a nearly step form, the index change Δn being about 6 %. The index difference between the substrate and the lens area is estimated with an accuracy of $\pm$ 3 %.

We have calculated Δn following the procedure suggested by HUGGINS and SUN /7/ with Ag_2O-dopant data given in /8/. Comparing

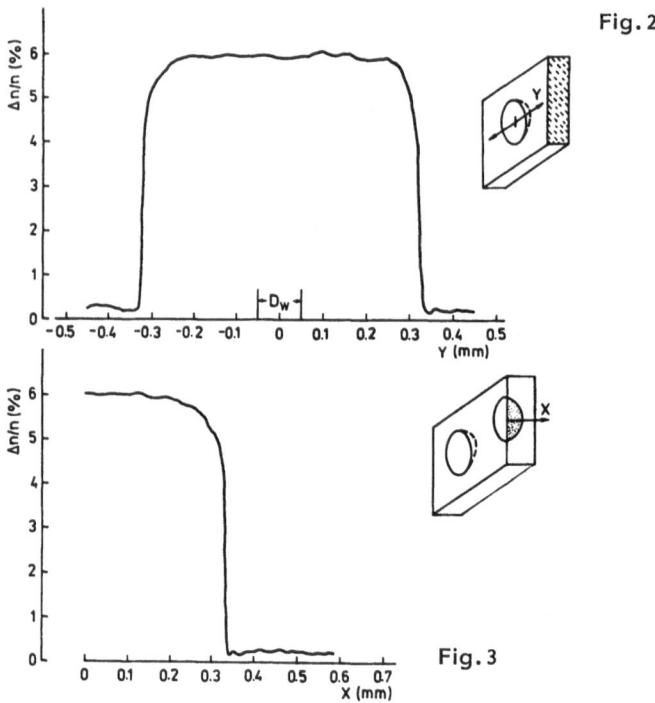

Fig. 2

Fig. 3

the measured and calculated values of Δn we conclude that an almost complete $Ag^+ - Na^+$ exchange occurs (94 %), i.e., the maximum index change for these type of glasses is reached.

As Δn is constant within the profile, the optical characteristics of the microlenses depend only on the value of Δn and the lens shape, i.e., the form of the silver ion front. In the case of a field-assisted ion exchange, the silver ion front is determined by the distribution of the applied electric drift field /9/. This means that the index profile can be calculated in terms of the technological parameters of the exchange process, such as D_w, migration time, temperature, voltage and so on.

We have observed that with the voltages applied, an almost hemispherical lens shape is produced when D_w is less than 100 μm. At $D_w > 100$ μm some deformations of the shape occur, the lens radius at the surface being greater than the ion penetration depth. These facts agree with the model presented in /9 /.

Assuming hemispherical microlenses and a constant Δn we have calculated the focal lengths f by the simple ray-tracing method.

The measured values of f are about 25 % less than the calculated ones. This discrepancy is assumed to be due to the approximations of a pure step-index profile and an ideal hemisphere. Our results show that more sophisticated lenses could be obtained using glasses in which higher Δn can be reached by silver-sodium ion exchange. For example, TiF_6 allows $\Delta n=0.22$ /10/. Sodalime glasses containing more than 20 % Na_2O are promising too.

4. Conclusions

A step-index distribution in planar arrayed microlenses is obtained by field-assisted ion exchange in glass. The optical characteristics of the lenses depend only on the lens geometry and the value of Δn. To improve the lens properties (NA, f), glasses are recommended which allow higher values of Δn to be reached by Ag^+- Na^+ exchange.

References

1. K. Iga, M. Oikawa, S. Misawa, J. Banno, Y. Kokubun: Appl. Opt. 21, 3456 (1982)
2. M. Oikawa, K. Iga, T. Sanada: Electron. Lett. 17, 452 (1981)
3. M. Oikawa, K. Iga: Appl. Opt. 21, 1052 (1982)
4. B.G. Panchev, I.T. Savatinova, A.G. Kebedjiev: Techn. Digest IEEE 1984, Florence, Int. Workshop on Integrat. Opt. Relat. Technol. for Sign. Processing, pp. 133-136
5. W. Eickhoff, E. Weidel: Opt. Quantum Electronics, 7,109(1974)
6. M. Ikeda, M. Tateda, H. Yoshikiyo: Appl. Opt. 14 814 (1975)
7. M.L. Huggins, K.-H. Sun: J. Am. Ceram. Soc. 26, 4 (1943)
8. S.D. Fantone: Appl. Opt. 22, 432 (1983)
9. H.-J. Lilienhof, E. Voges, D. Ritter, B. Pantschev: IEEE J. Quantum Electronics QE-18, 1877 (1982)
10. J.L. Coutaz, P.C. Jaussaud: Appl. Opt. 21, 1063 (1982)

Spherical Waveguide Multiplexing Device

V. Russo, S. Sottini, G.C. Righini, and S. Trigari
Istituto di Ricerca sulle Onde Elettromagnetiche del C.N.R.
I-Firenze, Italy

A wavelength multiplexer is described, characterized by a reflection grating coupled to a waveguide spherical geodesic lens. The device, particularly suitable for monomode fibers, can be used also as a tapping element.

The capacity of optical fiber transmission systems can be dramatically increased by multiplexing several signals at different wavelength on each fiber. Therefore an increasing interest exists for multiplexing-demultiplexing (mux-demux) devices. In general they consist of a dispersing element and a focusing one. Important characteristics are efficiency, reliability, cost, size and an easy coupling to input and output fibers.

In the case of multimode fibers a simple mux can be fabricated, for example, by a plane grating and GRIN optics, while for monomode fibers more complex devices have been proposed. In particular, Tomlinson suggested a waveguide geodesic lens as focusing element [1]. Such a lens has been recently realized by a simple printing technique [2], which satisfies the requirements of a low cost and simple technology. Unfortunately the

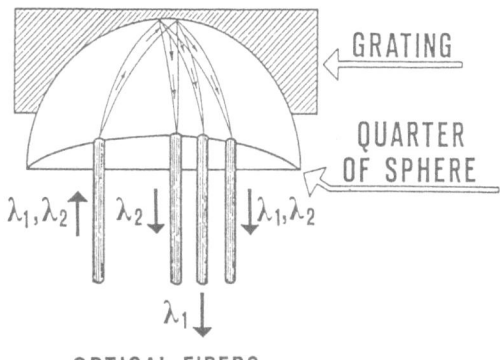

GRATING

QUARTER OF SPHERE

λ_1,λ_2↑ λ_2↓   ↓λ_1,λ_2

λ_1↓

OPTICAL FIBERS

Fig.1 – Mux-demux device utilizing an external grating coupled to a spherical waveguide geodesic lens. This device can be used also as a tapping component.

relatively high fabrication errors which can give rise to focal length shift and spherical aberrations limit the use of this technique to multi-mode fibers.

For this reason we have considered a mux-demux device based on the same working principle of the previous one but characterized by a spherical geodesic lens as focusing element, that is a waveguide laying on a quarter of sphere. As shown in Fig.1, the input and output fibers have to be coupled to the same edge of the thin film guide by butt joints secured by a transparent glue. The external reflection grating is positioned at the other lens edge, glued to the glass substrate.

In general, the 0th-order beam, reflected by the grating, is aberration free, as it can be easily proved by considering a hemispherical guide divided into two halves by a transmission grating and recalling its imaging properties. The wave aberration W of a diffracted beam can be evaluated with simple calculations and the usual approximations [3]. If Ω is its angle with respect to the 0th-order and α is the aperture, W turns out to be:

$$W = \Omega \, \alpha^3 / 6 \qquad (1)$$

It corresponds to the coma in bulk optics.

For example, let us consider a grating having 540 groves/mm, a lens with radius R = 10mm, and monomode fibers with NA = 0.1 corresponding to an aperture in the guide α = 0.067 (refractive index of the guide n=1.5). At λ = 0.633 μm, Ω turns out to be $\sim$ 0.35 rad and W = 0.18μm, that is a value very close to the Rayleigh's quarter wavelength rule and much smaller than the tolerance condition which is 0.6λ in the case of coma [4]. Moving to the infrared, at λ = 0.87 μm, Ω is $\sim$ 0.47 rad and W = 0.236μm = λ/3.68 while at λ = 1.3μm, Ω $\sim$ 0.7 rad and W = 0.35μm = λ/3.71. Therefore, in the case of monomode fibers a performance very near to the diffraction limit has to be expected. On the other hand, if we use graded index fibers with core diameter 50μm and NA=0.2, the beam aperture in the guide turns out to be α = 0.13. As a consequence, in the same conditions of the previous example, W = 1.28μm at λ = 0.633μm, W = 1.72μm at λ = 0.87μm and W=2.56μm at λ = 1.3μm. The wave aberration being larger than the wavelength, it is worth evaluating the transversal coma, given by $c_T = \Omega \alpha^2 / 2$. The resulting values of the coma are: c_T= 29μm, c_T= 39μm and c_T = 59μm respectively. If compared with the core diameter of the fiber, these results are reasonably good. Of course the coma reduces by using a wider grating pitch or fibers with a smaller NA.

The linear dispersion of the grating is given by:

$$R \, d\beta/d\lambda = Rm/\Lambda \cos \beta \qquad (2)$$

where m is the diffraction order, Λ is the grating pitch, and β is the angle of diffraction. A blazed grating is often convenient to minimize the

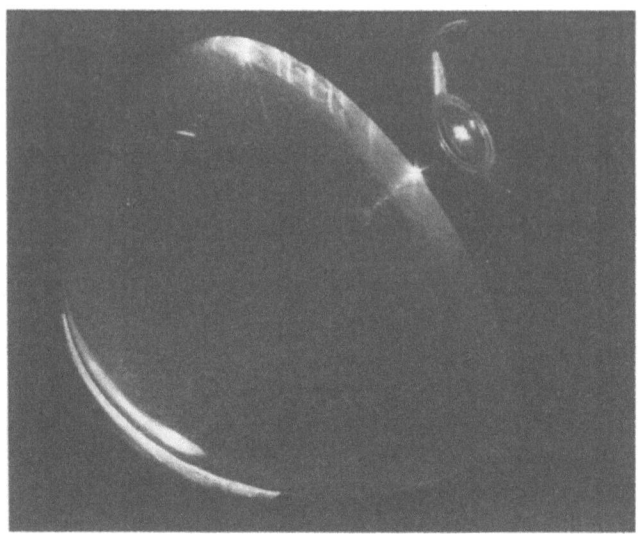

Fig.2 - Experimental set up utilizing a spherical waveguide made of epoxy
       resin deposited on a glass quarter of sphere.The quarter of
       sphere is laid and glued to a reflection grating visible in the
       bottom of the picture. The input light, from a microscope objec-
       tive, included two wavelengths, λ_1=0.633µm and λ_2=0.488µm. The
       bright lines at the lens edge are due to the light output from the
       film guide which is tapered.

losses. In our case the Littrow mounting of such a grating can be easily
achieved moving the input fiber along the circular edge of the guide so
avoiding the special cut at a slant angle of the lens substrate which is
necessary in the planar case. Of course eq.(1) is still valid with $\Omega = 2\phi$
where $\phi = \sin^{-1} \lambda/\Lambda$ is the blaze angle. From eq.(2), substituting β with ϕ
we have a linear dispersion:

$$R\, d\phi/d\lambda = R/\sqrt{\Lambda^2 - (\lambda/2)^2} \qquad (3)$$

With the same grating and geodesic lens of the example above described,
the values of the linear dispersion turn out to be: 5.48µm/nm, 5.55µm/nm
and 5.77µm/nm at $\lambda = 0.633$µm, 0.87µm and 1.3µm respectively. As a further
example, if a core spacing of 100µm is requested, the minimum channel separation $\Delta\lambda$ is 180Å at $\lambda = 0.87$µm.
    Preliminary experiments have been carried out by using a multimode
waveguide made of epoxy resin which was deposited on a glass quarter of
sphere with diameter 50mm. The input beam, coupled to the lens guide by a
microscope objective, Fig.2,included two wavelengths, λ_1= 6328Å (He-Ne laser) and λ_2=4880Å (argon laser). A reflection phase grating with 540

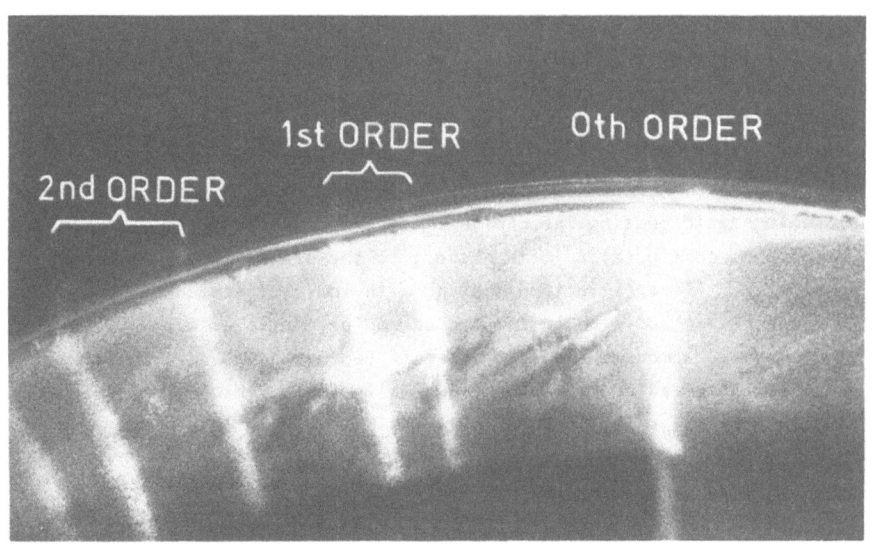

Fig.3 - Output edge of the spherical geodesic lens. The 0th, 1st and 2nd order are clearly visible.

lines/mm,glued at the lens output,reflected back a 0th-order beam and diffracted λ_1 and λ_2 beams into at least 2 orders as shown in Fig.3 where the output edge is imaged. The bright lines in the Figures are due to the light output from the film guide which is tapered at the edge. The 1st order λ_1 and λ_2 foci turned out to be ~ 2mm apart which is in good agreement with the theory. In the test, carried out with a spherical waveguide already available, it was impossible to get a measure of the insertion loss and the cross talk attenuation due to the disuniformity of the guide which was also tapered at the edges as already mentioned (Fig.3). Further experiments are in progress utilizing an ion-exchange waveguide and a blazed grating with 600 lines/mm.

In conclusion, the main advantage of the present approach with respect to the previous devices is given by the great accuracy - at low cost - available in the fabrication of the spherical geodesic lens so that the performance figured out by the theory should be actually achieved in experimental tests. Therefore this mux-demux device seems to be particularly interesting in the case of monomode fibers which are characterized by small core diameter and aperture angle. In practice, in this case the spherical geodesic lens can assure diffraction limited performance also with large diffraction angles.

With multimode fibers, having wider aperture, the aberration of the diffracted beams increase, but it is still acceptable for gratings up to 600 lines/mm. Moreover, current requirements for wavelength channel spacing are in the order of 40-60nm [5] and therefore gratings with period as low as 300 lines/mm are fully suitable: in this case the amount of coma is much lower than the fiber core diameter. If, on the other hand a higher dispersion is required and therefore gratings with frequency ≥600 lines/mm are to be used, the aberration of the lens becomes not negligible due to the high value of Ω, but it could be corrected anyway by designing a suitable holographic grating.The choice of spherical lens should also assure a more rugged configuration with respect to the planar one.

Finally, if a reflection grating with low diffraction efficiency is used, the device above described can be employed also as a signal tapping element where a monomode output fiber is fed by the aberration-free 0th order (Fig.1) reflected by the grating, while the diffracted monochromatic beams carry only a small fraction of the signal power.

References

[1] W.J.Tomlinson, U.S.Patent N.4, 153, 330 (1979)

[2] H.J.Lilienhof,E.Voges,D.Schulz, Proc.ECOC'82, p.321 (Cannes 1982)

[3] G.C.Righini,V.Russo,S.Sottini, AGARD Conf.Proc. n.219, p.25-1 (London 1977)

[4] M.Born,E.Wolf, Principles of Optics, (Pergamon Press 1980), chap.IX.

[5] G.Winzer, IEEE-J. Lightwave Techn. vol. LT-2, p.369-378, 1984

Sensor Applications of Low Finesse Integrated Optical Fabry-Perot Resonators

H. George, U. Hollenbach, J. Söchtig, and W. Sohler

Universität-GH-Paderborn, Fachbereich Physik, Angewandte Physik, Postfach 1621, D-4790 Paderborn, Fed. Rep. of Germany

Low finesse integrated optical Fabry-Perot resonators are presented as sensitive, quasi-digital sensors for a variety of physical parameters.

1) Introduction

Currently, fiber optic sensors are developed and investigated worldwide for the measurement of temperature, pressure, current, rotation rate etc. [1]. To achieve an all guided-wave sensor system, integrated optics is thought to develop optical circuits for signal processing applications [2]. Up to now, however, only few attempts have been reported to use integrated optical waveguides or circuits themselves for sensor applications. This is the more astonishing as integrated optical sensors would offer several advantages if compared with fiber optic devices: First, it seems to be easier to construct special waveguides, waveguide devices or even integrated optical circuits as sensors adapted to a given problem of measurement. Secondly, materials can be chosen, which are by far more sensitive to a change of a physical parameter than the corresponding fiber materials.

From the variety of possible integrated optical devices applicable as sensor elements, we decided to investigate low finesse integrated optical Fabry-Perot resonators (section 2). For the first time various sensor applications are demonstrated (section 4). Especially attractive is the possibility to use the resonators as quasi-digital devices. Appropriate methods of measurement were developed (section 3).

2) Sample Preparation

Low finesse integrated optical Fabry-Perot resonators (Fig. 1) were fabricated from conventional Ti:LiNbO$_3$ strip guides by carefully polishing their end faces perpendicular to the waveguide axis. The end faces form the resonator mirrors with a reflectivity of about 14 % due to the jump of the index of refraction from air to LiNbO$_3$. It was important to have monomode guides of low losses ($\lesssim$ 0.5 dB/cm) at the operation wavelengths (λ = 0.63 µm and 0.79 µm) to obtain definite resonances by a single mode alone.

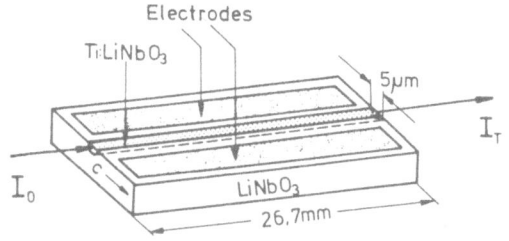

Fig. 1: Schematical drawing of a low finesse integrated optical Fabry-Perot resonator made of a Ti:LiNbO$_3$ strip guide

The length of the device of 26.7 mm was arbitrarily chosen; however, it determines the sensitivity as will be discussed below. Electrodes were prepared parallel to the waveguide; they allowed to demonstrate an electro-optical counting technique (next section).

3) Methods of Measurement

a) Resonator Characteristic

Integrated optical Fabry-Perot resonators have a periodic transmission and reflection as function of their optical path length. The lower the reflectivities of the resonator mirrors are, the more the resonances are smoothed. Nevertheless, the end face reflectivity of 14 % (together with low loss waveguides) is sufficient to produce pronounced structures. Corresponding results measured with our devices are presented as the upper curves of Fig. 2. In this case the optical path length was varied by the temperature of the resonator. The periodicity ΔT was 120 mK, determined by the wavelength λ, the temperature coefficients of length and of effective index of refraction and by the resonator length L itself:

$$\Delta T = \frac{\lambda}{2} \frac{1}{[\ L\ \partial n_{eff}/\partial T + n_{eff}\ \partial L/\partial T\ ]}$$

It is evident that especially the large coefficient $\partial n_{eff}/\partial T$ of LiNbO$_3$ (of the extraordinary index) is responsible for the high sensitivity of the device. As a consequence of the periodic characteristic, the number of orders of interference as function of a change of the optical path length of the resonator can be counted. To do this we have developed two different techniques.

b) Counting Orders of Interference

In principle, electronical counting of interference orders is easy by defining a threshold level for the photodiode signal to generate a pulse to

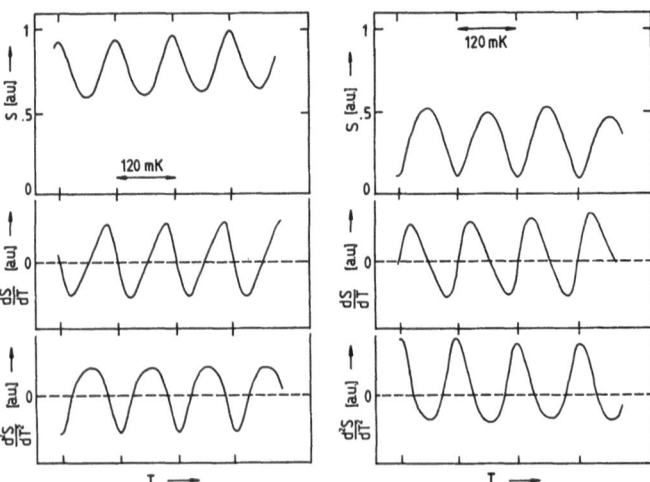

Fig. 2: Photodiode signals, undifferentiated (upper curves) and one respectively two times differentiated (middle and lower curves) versus the temperature. Left: transmission signals; right: reflection signals (see also Fig. 3)

be counted (two pulses within each order). However, the determination of the direction of a temperature change e.g. is not possible. It requires a second - if possible identical - signal shifted in phase at $\pi/2$ with respect to the first. Then a simple logic using both signals can decide if the corresponding pulse has to be added or subtracted in the counter.

In our experiments we did not use the original signals for counting, but the one and two times differentiated signals. They are phase-shifted with respect to one another and oscillate around zero due to the differentiation (see Fig. 2). That allows to choose zero as the trigger point and to become in this way independent from slow changes of the light intensity.

i) Differentiation by Electrooptical Modulation. Differentiating the signals S (in reflection or transmission) was achieved by applying a small ac-voltage of frequency f ($\sim$ 1 kHz) at the electrodes of the resonator (Fig. 3). In this way the optical path length respectively the optical phase difference $\phi = 4\pi n_{eff}L/\lambda$ between the interfering waves inside the resonator was electrooptically modulated via the index of refraction. By measuring the Fourier components of the photodiode signal at the frequencies f and 2 f with phase sensitive amplifiers ("Lock-in"), the first respectively second derivative of the original signal were obtained. This is easily understood by looking at a Taylor expansion of the signal:

$$S(\phi_0 + \Delta\phi) = S(\phi_0) + \frac{dS}{d\phi}\Big|_{\phi_0} \Delta\phi + \frac{d^2S}{d\phi^2}\Big|_{\phi_0} (\Delta\phi)^2 + \ldots$$

$\Delta\phi$ varies with f, $(\Delta\phi)^2$ has a term varying with 2f. The measured Fourier components therefore are proportional to $dS/d\phi$ and to $d^2S/d\phi^2$, which in turn are proportional to dS/dT and to d^2S/dT^2 in the case of our example of Fig. 2.

ii) Differentiation by Wavelength Modulation. An equivalent modulation of the phase ϕ can be obtained via a wavelength modulation of the laser light. This allows to use the integrated optical sensor without electrical leads, only supplied with light by an optical fiber. This is by far the more favourable method as an all optical, guided wave approach.

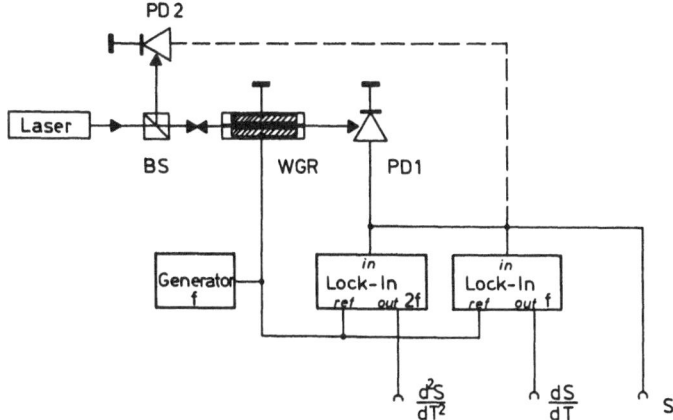

Fig. 3: Schematical experimental arrangement to differentiate the photodiode signals by electrooptical modulation.

Wavelength modulation was done with a semiconductor laser (λ = 790 nm) via a current modulation (see Fig. 4). Precondition was a good frequency stabilization of the (transversal and longitudinal monomode) laser. This required a stabilized dc-current source and - more difficult - an extremely good temperature stabilization of the laser itself. With a carefully designed package including a peltier-cooler and with appropriate electronics,a temperature stability of about 1 mK was achieved (see lower inset of Fig. 4); this figure corresponds to a - calculated - frequency stability of 24 MHz. The current modulation resulted in a frequency modulation of an amplitude of $\Delta\nu/2$=250 MHz, observed with a scanning (bulk) Fabry-Perot (see right inset of Fig. 4). It generated a phase modulation $\Delta\phi \sim 0.6$ rad inside the resonator allowing to apply the counting technique in exactly the same manner as described above.

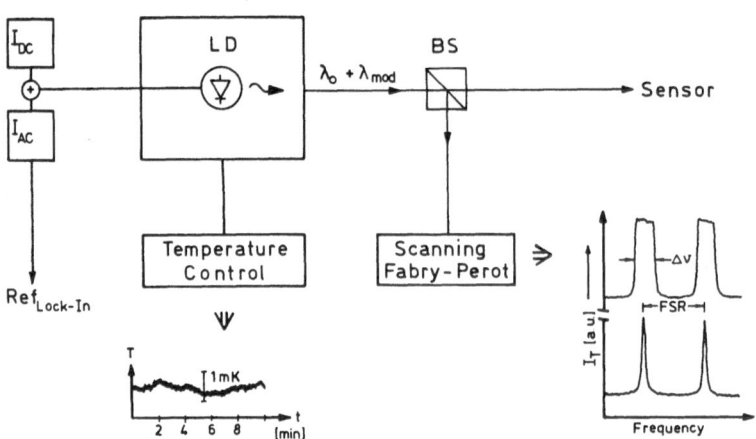

Fig. 4: Schematical experimental setup for laser diode frequency stabilization and modulation. Lower inset: laser temperature versus time; Right inset: scanning Fabry-Perot spectra of the light of the modulated (upper curve) and unmodulated laser (lower curve). Free spectral range: 1.8GHz.

4) Applications and Experimental Results

Results of the counting technique are presented in Fig. 5 as the number of pulses versus a temperature change. In this example four pulses per order of interference were generated (at every zero of the differentiated signals of Fig. 2) and counted, giving a resolution of approximately 30 mK/pulse. The optical path length of the resonator is not only dependent on the temperature as seen by the example of Figs. 2 and 5, but also on uniaxial pressure (force)

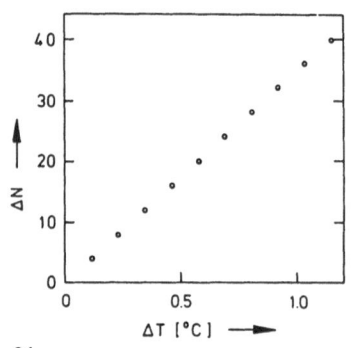

Fig. 5: Number of pulses ΔN from an integrated optical sensor-resonator as function of a temperature change ΔT

applied parallel to the c-axis or parallel to the waveguide axis, on voltage (respectively electric field strength detected by an antenna) applied to the electrodes and on the refractive index of the medium covering the resonator. Therefore,a number of further applications result (which are to be discussed in more detail at the conference). However, due to the dependence on various physical parameters, care has to be taken to measure the influence of on parameter alone. A possible solution of this problem is e.g. the combination of a sensor-resonator with a reference device. Furthermore, as only changes of the optical path length can be measured in the described manner (and not its absolute value), appropriate applications have to be chosen.

5) Conclusion

Low finesse integrated optical Fabry-Perot resonators have been fabricated and presented as sensitive sensors for various physical parameters. Their periodic characteristic allowed to use them as quasi-digital sensors; appropriate counting techniques were developed. Supplied with light via an optical fiber, an all guided wave sensor system with its well known advantages was demonstrated.

References

1   e.g. Proc. 2nd Int. Conf. on Optical Fiber Sensors, Sept. 5 - 7, 1984, Stuttgart, VDE-Verlag GmbH, Berlin

2   e.g. H.J. Arditty, J.P. Bettini, Y. Bourbin, Ph. Graindorge, H.C. Lefèvre, M. Papuchon and S. Vatoux in Ref. 1, p. 321

Fiber Optic Mach-Zehnder Interferometer Based on Lithium Niobate Components

P.O. Andersson, G. Edwall, A. Persson, and L. Thylén

Telefonaktiebolaget L M Ericsson, Fibre Optics and Line Transmission
S-126 25 Stockholm, Sweden

An integrated optics version of a Mach-Zehnder interferometer
is presented. Sensitivities of the same order as with other Mach-
Zehnder concepts were reached.

Fiberoptic Mach-Zehnder interferometers are today of great interest in
many sensor applications. These sensors offer high sensitivity to several types of fields, including: magnetic and electric fields, temperature, acceleration, sound and strain. In order to get a viable sensor
concept that works outside the laboratory environment the packaging is
crucial. In general, two main routes are followed to accomplish an acceptable sensor, the all-fiber interferometer and the integrated optics
version (1). Most Mach-Zehnder interferometers presented today are of
the all-fiber type using spliced fiber couplers. Usually electro-mechani-
cal phase-tracking components are used,which are bulky and suffer from
a restricted linear frequency range.

Both splitting, recombining and phase compensation are easily accomplished in $LiNbO_3$ integrated optics devices. The light source and the detector can be hybrid-integrated to the $LiNbO_3$ chip, leaving only the fiber
of the sensing and reference arm outside the chip.

Experiments

We have fabricated a Mach-Zehnder type interferometer with passive Y-
type splitters and combiners and active phase modulators in the same
chip. The principal design is shown in fig 1. A 1.3 μm single mode laser
diode was chosen as a lightsource in order to minimize the photo-refrac-

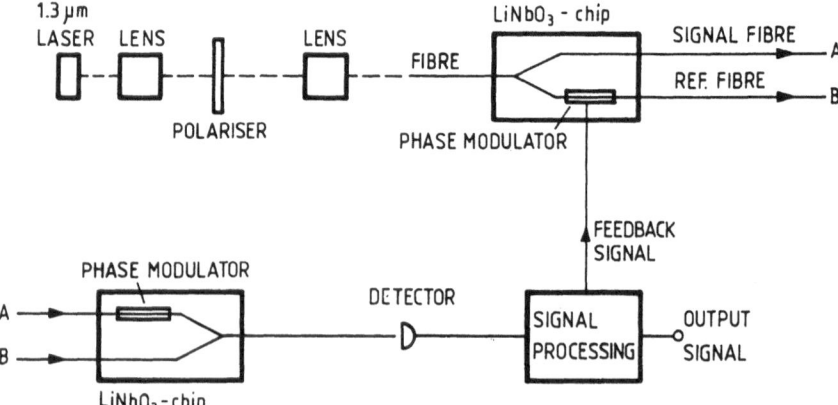

Fig 1   Experimental set-up of the interferometer using integrated optics

tive effects in the LiNbO$_3$. The laser was operated at a current of 77 mA at 25°C, resulting in a laser effect of 3 mW.

Polarization preserving fibres were used throughout. This will enhance the sensitivity of the interferometer due to restricted coupling between different polarization states in the fibre. Care has been taken to keep the length of the two fibres constituting the interferometer arms equal to within at least one millimeter. This is important, since the phase noise of the laser will increase the noise level in the interferometer directly proportional to the length difference between the two interferometer arms. The detector was an InGaAsP PIN diode.

The layout of the LiNbO$_3$ device is shown in Fig 2. Ti indiffused channel waveguides in Z-cut Y-propagation LiNbO$_3$ were used. In the fabrication process of the waveguides we used a 600 Å Ti layer and 16 hours diffusion time at 980°C in H$_2$O-rich atmosphere. The SiO$_2$ buffer thickness was approx 600 Å. We used Aluminium electrodes and lift-off for both Titanium and Aluminium. The endfaces were polished and the fibres buttcoupled to the LiNbO$_3$ chip. The insertion loss for one Y-branch including coupling losses at the endfaces and without optimization for the specific fibre used was 6 dB. The phase-shifting capability for one phase modulator was 0,3 π /V/cm.

At this stage, a flexible design was chosen with all components easily exchangable. The fibers were aligned and oriented in silicon V-grooves before butt-end-coupling to the waveguides in the LiNbO$_3$(Ti), fig 3.

PHASE MODULATORS

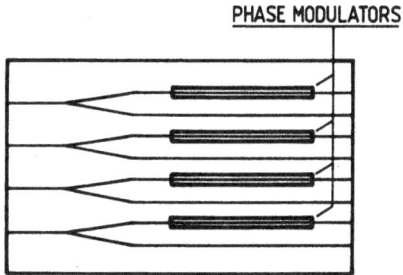

LiNbO$_3$ - DEVICE

Fig 2   Layout of the LiNbO$_3$-device. Physical length, 27.0 mm. Active length of phase-modulators, 13.5 mm. Split angle of Y-branch, 1.2 degrees.

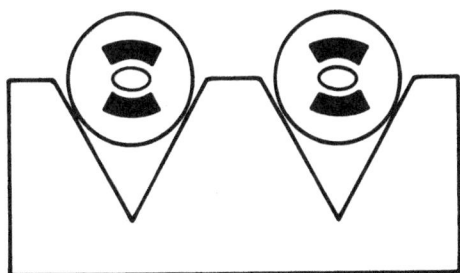

Fig 3   Fibres in silicon V-grooves. The bow-tie structure is easily seen.

Results and Discussion

The sensitivity of the interferometer was 6 μ rad/$\sqrt{Hz}$, the visibility −5 dB, and the signal to noise ratio 25 dB at 30 kHz bandwidth. The frequency response was linear from dc to 30 kHz-limited by the detector electronics. The total loss was approx 33 dB, which is relatively high and caused by the butt-coupling technique used. The concentricity of the fibre used is not sufficient to simultanously align both fibres to the waveguides without increasing the loss. Still,we used this technique because it offers greater flexibility. Both loss and visibility will be improved using the coupling technique described in ref (2). Potentially,the interferometer could be used up to several GHz by increasing the bandwidth of the detecting system, (3).

With the detection scheme used, the signal from the detector will be;

$$I = A^2 + B^2 + 2*A*B*\cos \varphi \qquad (1)$$

where A and B are the amplitudes of the light in the two interferometer-arms respectively and φ is the phaseshift induced by the signal.

A drawback of this design is that quadrature will not be maintained if the intensities in the two interferometerarms varies, as the detected intensity in quadrature corresponds to;

$$I = A^2 + B^2 \qquad (2)$$

In this interferometer the variations depended on altered coupling-efficiency from the laser to the fibre and from the fibre to the $LiNbO_3$-devices due to temperature variations in the translators used. This will not be the case in a fieldversion of the interferometer using epoxy to fix the fibre to the $LiNbO_3$ chip. A more serious case is the microbend loss that will appear when bending the two interferometer arms in an outdoor environment. This problem can be eliminated if a two-detector scheme with the detectors organized in a "push-pull" arrangement is used.

To summarize, a first generation of a lithium niobate-based Mach-Zehnder fiber optic interferometer has shown sensitivities of the same order of magnitude as other Mach-Zehnder concepts. With improved design of the $LiNbO_3$ devices,still better results are anticipated. The integrated-optics type of interferometer lends itself to small and rugged packaging of the sensor and a simple detection system.

References

1   T G Giallorenzi, et al: "Optical Fiber Sensor Technology" IEEE J Quantum Electronics, QE-18, 626-665 (1982)

2   K H Cameron, "Simple and Practical Technique for Attaching Single-Mode Fibres to Lithium Niobate Waveguides", Electron Lett, NO 23, 1984, pp 974-976

3   L Thylén et al: "Computer analysis and design of Ti: $LiNbO_3$ Integrated Optics devices and comparison with experiments", Proc ECOC 83, Elsevier, Amsterdam (1983), 425-428

Part II

Material and Fabrication

Applications of Electron Beam Lithography to Integrated Optics

C.D.W. Wilkinson

Department of Electronics and Electrical Engineering, University of Glasgow
Glasgow, G12 8QQ, United Kingdom

Lithographic technology is vital to integrated optics as it is the means of defining the pattern of stripe waveguides. In the wavelength range 0.5 to 1.5 microns, stripe guides have widths between 2 and 8 microns and to ensure low scattering losses must have very smooth edges ($0.1\mu m$ or better for spatial frequencies between 0.1 and $10\mu m^{-1}$). Compared with patterns used in integrated electronic circuits, the ones employed in integrated optics are sparse, have very high aspect ratios and are not wholly rectilinear; tapers and bends are widely used.

Almost all automatic pattern generators (whether used with optical or electron beam lithography) employ plotting elements defined in rectangular coordinates and thus tapers and bends must be composed of an assembly of suitable small rectangles. Given the extreme edge smoothness required this implies that the smallest rectangle must be a very small fraction of a micron.

Thus pattern generation for I/O masks is mainly done either using non-automatic methods to draw the patterns (a knife on a table which can be rotated) or by electron beam lithography using a very small picture element (pixel) size. The advantage of direct optical methods lies in their relative cheapness whereas the advantage of electron beam methods lies in the direct translation of readily changed designs into a final pattern.

While in optical methods of pattern generation it is normal to make a mask and print using U.V. light onto photoresist on the optical substrate, in electron beam lithography direct writing of the pattern onto the substrate is possible. If the details are smaller than the resolution of U.V. printing (about 0.3 micron at best) direct writing is essential for arbitary patterns. In principle X-ray printing of masks written by electron beam lithography can achieve a resolution of at least 50nm.

Sophisticated software programs have been written to ease the task of the I/O designer and ensure automatically edge smoothness.

An interesting test case for optical versus electron beam lithographic methods arises in the fabrication of grating reflectors for semi-conductor lasers. The grating pitch must be a half of the effective wavelength – which implies a pitch of $0.12\mu m$ for GaAs devices and $0.2\mu m$ for $1.55\mu m$ InGaAsP devices. Not only is the required linewidth in the 0.05 to $0.1\mu m$ range, the placing of the lines must be such that no significant phase errors are introduced which implies that the lines be placed to better than a quarter of the pitch. Such gratings can be made by the exposure of photo-resist to crossed beams of U.V. laser light which gives excellent pitch and spacing accuracy; however reproducibility of the grating depth is very hard to achieve.

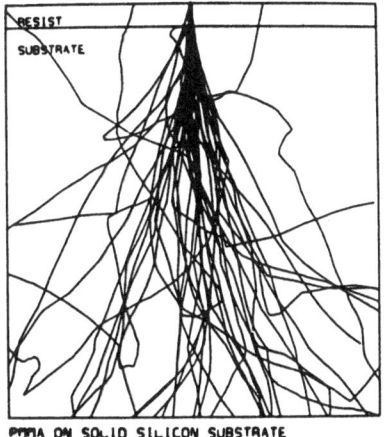

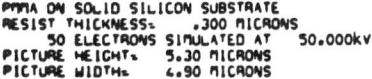

PMMA ON SOLID SILICON SUBSTRATE
RESIST THICKNESS: .300 MICRONS
    50 ELECTRONS SIMULATED AT   50.000kV
PICTURE HEIGHT: 5.30 MICRONS
PICTURE WIDTH: 4.90 MICRONS

Fig 1: Monte Carlo Simulation
of a 50kV point beam.

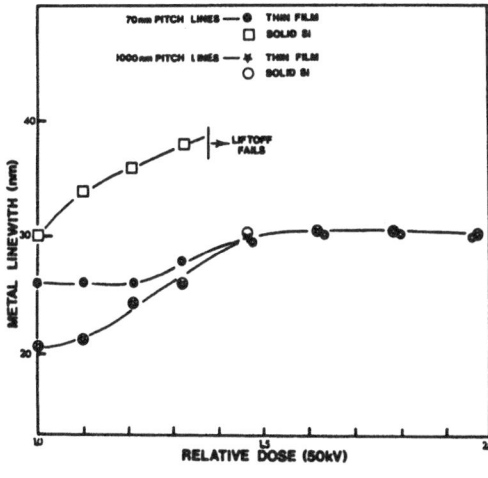

Fig. 2: Linewidth of metal lines as a
function of relative exposure dose
on thick and thin substrates.

The pitch requirements for first order gratings are difficult to achieve
using convential electron fabrication machines - but 1.55μm lasers incor-
porating second order (0.4μm pitch) made by electron beam lithography have
been demonstrated (1).

To see whether it is possible to make first order gratings using electron
beam lithography, consider the interaction of a high voltage electron with
a polymeric resist on a solid substrate. Figure 1 shows a Monte Carlo
simulation of fifty 50kV electrons striking 0.3μm of resist on a Si substrate;
notice the spreading inside the resist and that some electrons return to the
resist at considerable distances from the point of entry. This suggests
that the highest resolution will be obtained with very thin resist, prefer-
ably on a very thin substrate (membrane). Figure 2 shows a comparison of
the linewidth achieved using a two level PMMA resist of total thickness 60nm
on a 60nm Si_3N_4 (membranes and on solid Silicon) (2). It can be seen that
the minimum linewidth of some 26nm does not vary much as a function of
exposure dose on thin and thick substrates if the lines are well separated;
but for closely spaced lines (60nm pitch), the exposure range on a thick
substrate is very limited. 50nm lines on a 100nm pitch (with linewidth
accuracy of 2.5%) have been made on membranes; to achieve this on solid
substrates will require an electron beam size of 10nm or less (much smaller
than that used in conventional rather than high resolution electron beam
machines) and a high voltage (⪰50kV) (3). It is interesting to note that
Si_3N_4 membranes can be used as masks for printing by soft X-rays.

A direct use of electron beam lithography lies in the fabrication of
metal clad stripe optical waveguides. Devices (switches, directional
couplers, etc.) made using all dielectric waveguides are much longer
(> 1000λ) than the equivalent metal clad microwave devices (< 5 λ or less).
Moreover certain devices (steerable optical antenna arrays with the
necessary element spacing of 0.5λ or less) are only possible in metal clad
guides. The attenuation in such guides is high, but acceptable for short

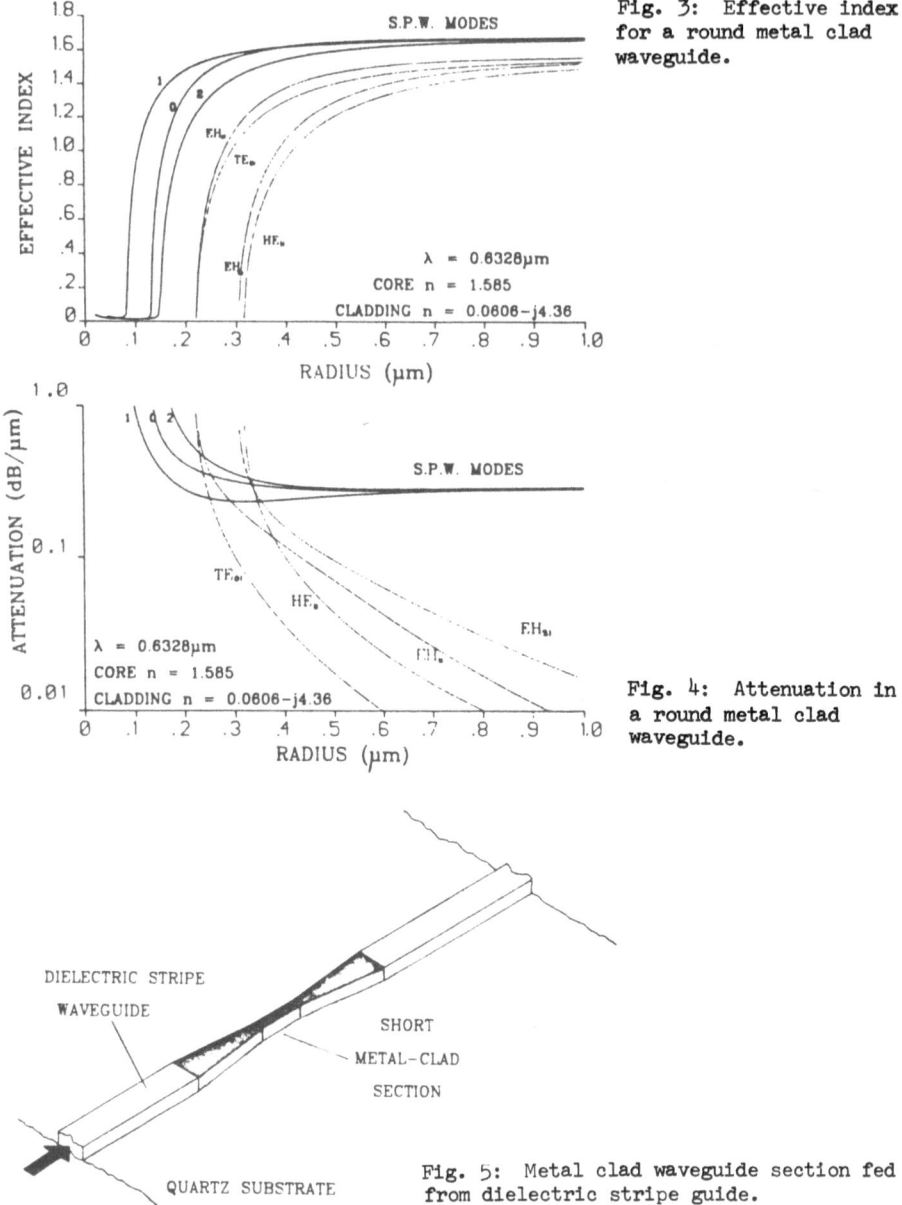

Fig. 3: Effective index for a round metal clad waveguide.

Fig. 4: Attenuation in a round metal clad waveguide.

Fig. 5: Metal clad waveguide section fed from dielectric stripe guide.

devices. Figure 3 shows the modes of a round metal clad optical waveguide at a wavelength of 0.63μm and Figure 4 the corresponding attenuation. Such waveguides have been made, using electron beam lithography and reactive ion beam etching in an oxygen plasma and tested by coupling light from a dielectric waveguide into a metal clad section (see figure 5). With sufficiently small guides cut-off was observed and overall transmission losses of the order of 10 dB over a 40μm length of guide were observed.

Acknowledgements

The work of many of my colleagues in the Department of Electronics and Electrical Engineering at the University of Glasgow is discussed in this paper: my thanks go to Dr. S.P.Beaumont, Dr. S.Mackie, Mr. C.Binnie, Mr. C.MacGregor and Mr. J.Crichton. This work is supported by the Science and Engineering Research Council (United Kingdom).

References

1) L.D.Westbrook, A.W.Nelson and C.Dix, Electronics Letters 10,423 (1983)
2) S.P.Beaumont, B.Singh and C.D.W.Wilkinson, American Electrochemical Society Meeting, Montreal, Canada (1982)
3) S.P.Beaumont, P.G.Baver, T.Tamamure and C.D.W.Wilkinson, Appl. Phys. Letts. 38, 436 (1981)

Laser Beam Photolithographic System for Integrated Optics Applications

I. Ben-David, S. Berlowitz, M. Itzkowitz, S. Ruschin, and N. Croitoru

Department of Electron Devices & Materials & Electromagnetic Radiation
Tel-Aviv University, Ramat-Aviv 69 978, Israel

A photolithographic system was operated based on direct laser beam explosure of photoresist films. Lineshape dependence on various fabrication parameters is reported.

Introduction

Integrated Optic devices, like channel waveguides and directional couplers, have special properties as compared with conventional microelectronic structures. These properties are a high aspect ratio (ratio between length and width) of the generated patterns and, at the present stage of I.O. technology, a relatively low density of information contained in the circuit layout. The system presented here may provide a simple low-cost answer for most Integrated Optic applications of current interest.

In the present system, the substrate on which the desired pattern is to be generated is coated with a photoresist film which is directly exposed to a focused laser beam. The pattern is then generated by displacing the substrate, which is placed on a couple of accurate x-y positioners. The carriage position is controlled by means of a microcomputer, which has stored in its memory the shape of the pattern. Special software was programmed in order to facilitate the design steps. The system has the advantage of simplicity and fabrication time reduction, when compared to conventional microphotographic systems. When compared to other directly-inscribing techniques, such as electron beam or X-ray lithography, this system has the advantage of a significantly reduced cost. The present system has been successfully applied by us also to the fabrication of SAW electrode transducers of special shapes. The application of a laser beam pattern generator for reticle and mask generation of VLSI circuits has been also recently reported (1). In the following, the main parts of the system will be briefly described, and experimental results will then be given, relating the dependence of the line width and shape on various fabrication parameters. So far, line patterns on photoresist have been obtained of widths down to 1 µm with edge definition of about 0.1 µm.

System Description

The system is composed of five parts (see Fig.1): a laser, a microcomputer, an optical focusing system, an x-y positioning system, and a laser intensity control system.

a) *The Laser*. We use here a He Cd laser irradiating at a wavelength of 442 nm and a continuous power of 5 mW. This wavelength was chosen since it is still in the visible region, a fact that facilitates the construction and alignment of the optical system. The wavelength is also within the range of conventional commercial photoresists, such as AZ 1350J.

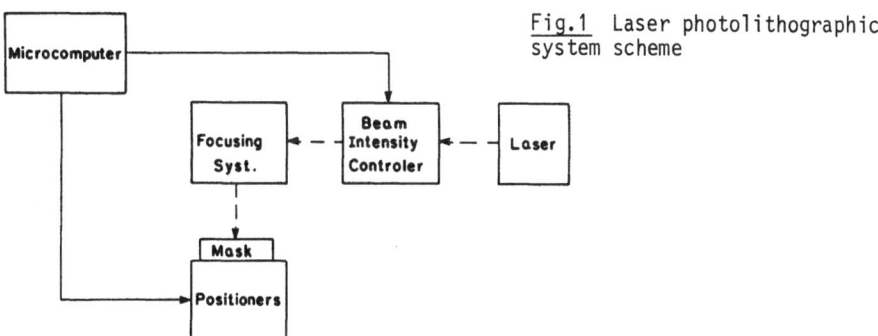

Fig.1 Laser photolithographic system scheme

b) *The Microcomputer*. An Apple II-e microcomputer is used which has the following roles: (1) Preparation and design of the mask pattern and its storage on a disk; (2) Process control. The microcomputer controls the positioning and intensity of the beam by means of an 8-bit binary word.

c) *The x-y Positioners*. The x-y positioners are composed of two independent tables driven by step motors of 0.1 micron steps. The tables are place perpendicularly to each other, allowing the generation of straight, diagonal or any desired lineshape.

d) *The Optical Focusing System*. The optical focusing system consists of a series of lenses, that are interchangeable according to the desired spot size. The system was designed in such a way that the focusing condition can be checked on real time.

e) *The Beam Intensity Control System*. The beam intensity control system is composed of an acousto-optic modulator and a closed-loop digital control system.

Experimental Results

A systematic study was performed for the dependence of the line width and shape on various fabrication parameters. The photoresist used was Shipley AZ 13505, mixed in various ratios with Microposit Thinner. The parameters studied that affected the line fabrication were: photoresist concentration, photoresist spinning rate, development time, and irradiation intensity. The beam scanning velocity was kept fixed in all the present measurements and was of 0.1 mm/sec.

In Figure 2, we see the linewidth as a function of the photoresist thickness. The thickness was controlled by varying the resist concentration and spinning rate. The intensities marked on the graph lines were measured after back reflection from the sample as explained below. We observe here a clear dependence of the linewidth on the resist thickness. A thicker resist layer produced a narrower line. The linewidth defined here is the net width of substrate exposed after development.

Figure 3 shows the dependence of the linewidth on the laser irradiance intensity. The intensity here was measured in real time in the back reflected beam outgoing from the sample. The back reflection was also used as a means of controlling the position of the sample in the focal plane of the focusing system. The absolute intensity on the focal spot was measured to be of the order of 1 µW after significant attenuation. The different

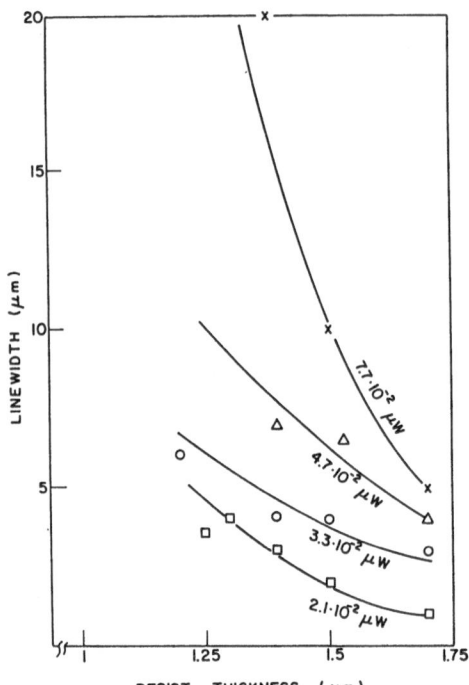

Fig.2 Linewidth as a function of the photoresist thickness for various irradiation intensities

Fig.3 Linewidth as a function of laser irradiation intensities for different spinning rates and resist dilution ratios

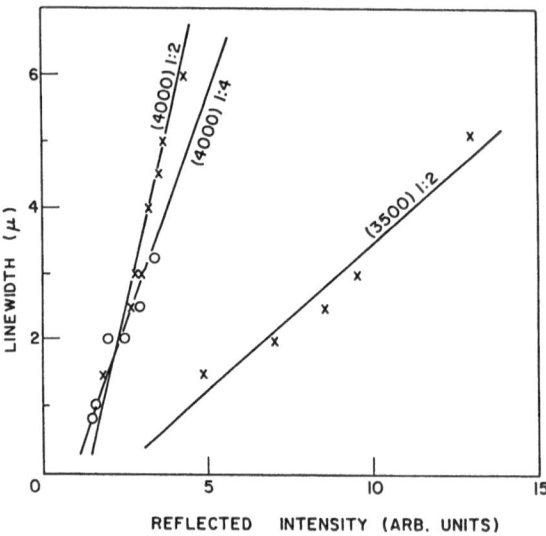

curves on the graph correspond to different resist thicknesses, obtained by varying the dilution ratios and spinning rates.

A more detailed view of a typical channel is shown in Fig.4, where an overview of the channel was photographed by means of a Scanning Electron

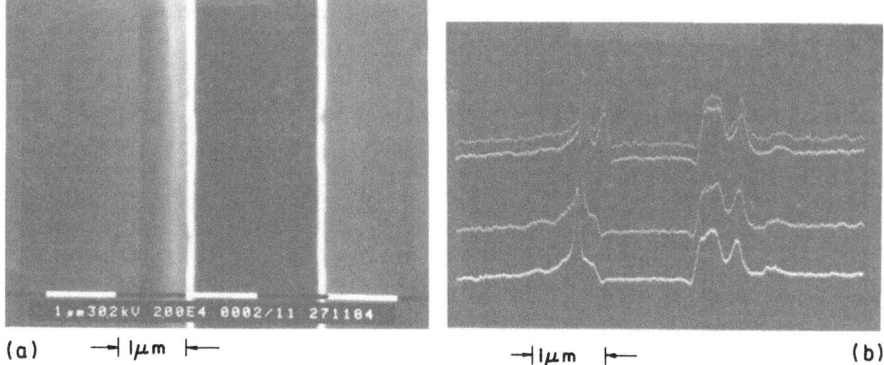

(a) ⊣ 1μm ⊢   ⊣1μm ⊢   (b)

<u>Fig.4</u> Scanning electron micrographs of channels on photoresist: (a) channel
overview; (b) "y mode" intensity scan

Microscope. The white lines show sharp edge definition of the order of
0.1 μm. The photography shows also additional fringes originated by diffraction effects. A special effort is being invested now on reducing the
fringing effects. There is a marked asymmetry in the fringes, an effect
which is presently under theoretical investigation. The fringing effect is
also apparent from Fig.4b, where a "y-mode" scanning of secondary electron
emission intensity was performed.

Conclusions

The system described seems to be able to produce Integrated Optical devices
more efficiently than the standard VLSI photographic techniques and at a
more reduced cost as compared to electron beam lithography. The pattern is
directly created on the substrate, and pattern variations are easily performed by means of software. The intensity control enables continuous linewidth changing during the run. This ability of having several intensity
levels on the same patterns is a useful feature of this system, and can be
applied to detailed research of the laser photoresist interaction in order
to fabricate 2-D and 3-D structures in the photoresist. The best resolution
obtained so far, after developing, was of 0.8 μm, corresponding to 1.3 L(th),
where L(th) is the theoretical resolution defined by $\lambda/2(NA) = 0.6$ μm.

References

1. Mac Donald, D.B., Nagler, M., Van Peski, C., and Whitney, T.R., 1984,
   Proc. SPIE. 470, 212-220

Laser Induced Chemical Vapor Deposition of Metals for Electrical Contacts in Semiconductor Processing

D. Braichotte and H. van den Bergh

Institut de Chimie Physique, Ecole Polytechnique Fédérale (ETH)
CH-1015 Lausanne, Switzerland

Pyrolytic and photolytic laser chemical vapor deposition of copper, platinum, gallium and tin is studied for the production of low resistance ohmic contacts of very small dimensions.

1. Introduction

Metallic contacts for integrated circuits and optoelectronic devices are usually produced by a multistage photolithographic process. More recently focussed laser beams have been used to deposit well localised shapes of metallic and other materials, on many kinds of substrates. Lasers have also been used for the photochemical etching of small areas on semiconductor surfaces. The two latter subjects have been reviewed recently (1-6).

The present paper concentrates on some new results obtained in the photolytic and pyrolytic laser-induced chemical vapor deposition (LCVD), in which organometallic compounds are used to produce microscopical electrical contacts. Contrary to photolithography, the laser direct writing method used here is a single stage process, which in part due to the use of metal-organic compounds can be integrated easily with chemical vapor deposition techniques used for instance in the production of III-V optoelectronic devices.

One of the contributions from our laboratory to this field of research was the introduction of the hexafluoroacetylacetonate derivatives of several metals for use in LCVD (5,7,8), and several laboratories have more recently also followed this way (9-12). Below, some results are presented on the laser direct writing with copper and platinum, together with some results on tin. The latter may be of use for ohmic contacts to n- and p-type indium phosphide (13).

2. Experimental

The experimental setup has been described in some detail previously (7,8, 14), and only some of the more essential features are briefly outlined here. The electrical resistance is measured by essentially the same method as described in reference 15, and 4 point measurements will be available in the near future. For transmission and diffractive electron microscopy the metallic films are deposited on a thin layer of amorphous carbon of about 100 Å thick. These carbon layers are made by vacuum evaporation, and then transposed onto a fine copper wire grid. Clearly, the structure of the deposited metal film and the deposition rate, as well as the chemical purity and resistivity, will depend on the substrate, as will be discussed below. Still, a lot of useful information on pyrolytic and photolytic LCVD can be obtained from these depositions on

carbon. The measurements of the resistivity presented here are on pyrex
substrates for the pyrolytic depositions and on fused silica for the
photolytic depositions. Pyrolytic LCVD is done with the green line of the
Ar ion laser at 514.5 nm. The reported laser beam power densities are to
be considered as upper limits as they are calculated from the diameter of
the laser beam at its focal point for the diffraction limited case:
$2W_0=1.27\lambda f/d$, where $2W_0$ is the beam diameter at the focal point, λ is the
wavelength, f the focal length of the lens, and d the diameter of the
laser beam prior to focussing. The actual power densities may well be as
much as a factor of two lower, and in any case due to the Gaussian intensity distribution in the single mode laser beam the reported densities
are only averaged values. Photolytic LCVD is done at 257 nm with the frequency doubled green light of the Ar ion laser. It has been noted before that
in some cases of LCVD one does not have either pure pyrolytic or pure photolytic LCVD. In this work we believe that in the case of photolytic LCVD
the contribution of surface heating is small or negligible (of the order
of 1 K or less) whereas in the case of pyrolytic LCVD the contribution
of photolysis at 514.5 nm is also extremely small or negligible. Most
chemical compounds used certainly have small extiction coefficients at
this wavelength, but their photochemistry with green light has generally
not been investigated.

The materials used are platinumbishexafluoroacetylacetonate (PtHFAcAc)
synthesized and purified following reference 16, copperbishexafluoroacetyl-
acetonate synthesized and purified according to reference 17 (CuHFAcAc),
tintetramethyl (TMSn) from Ventron (99.5%) which is thoroughly degassed
prior to use, and an inert gas, generally He (Carba Gaz N46).

3. Copper deposition

Fig. 1 shows the absorption spectrum of gaseous CuHFAcAc at 60 °C where
the vapor pressure is about 0.2 torr. The decadic molar extinction coefficient of this substance at wavelengths of interest for photolytic LCVD, i.e.
between 190 and 350 nm, are of the order of $1-2\times10^4$ l mole^{-1} cm^{-1} which is
quite appreciable. It is realised that high absorption coefficients do not

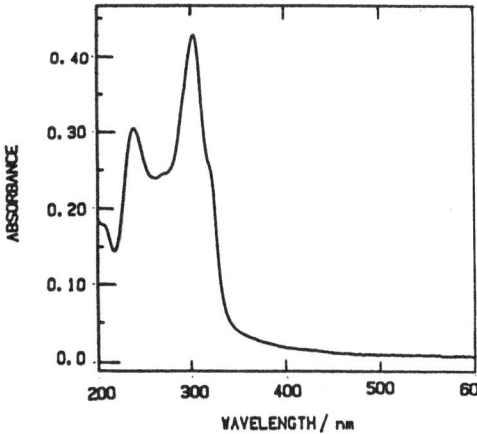

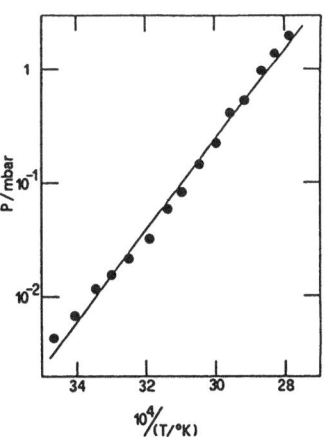

Fig. 1. Absorption spectrum of Copperbis-
hexafluoroacetylacetonate in the gas phase
at 60 °C.

Fig. 2. Plot of the vapor
pressure of Cu(HFAcAc)$_2$ against
the inverse temperature.

necessarily imply efficient photodissociation to yield free copper atoms, either on the surface or in the gas phase. Furthermore the spectrum of the chemisorbed molecule which we are presently studying can be shifted significantly from the gas phase spectrum. Fig. 2 shows the dependence of the CuHFAcAc vapor pressure on the inverse temperature. From the slope of this curve the heat of sublimation of this substance can be found for the temperature range of the measurements. We find ΔH(sublim.,<T>) to be 18.5 kcal mole^{-1}.

Fig. 2 shows a typical result as seen in a scanning electron microscope of a copper ohmic contact deposited by pyrolytic LCVD on glass. The scale can be seen from the measure at the bottom left of the picture,which indicates a real length of 100 microns. The temperature of the cell containing the CuHFAcAc is 90 ^{o}C, and the writing speed which is obtained by moving the target while keeping the laser beam stationary is 17 microns per second. The laser power density at 514.5 nm is 60 kW cm^{-2}. It is demonstrated that at this power density at the applied conditions the ohmic contact is quite regular. The starting point and the terminal point are exposed for the same conditions,but now with the target stationary. This gives rise to somewhat more developed deposits, the nature of which is crystallin as can

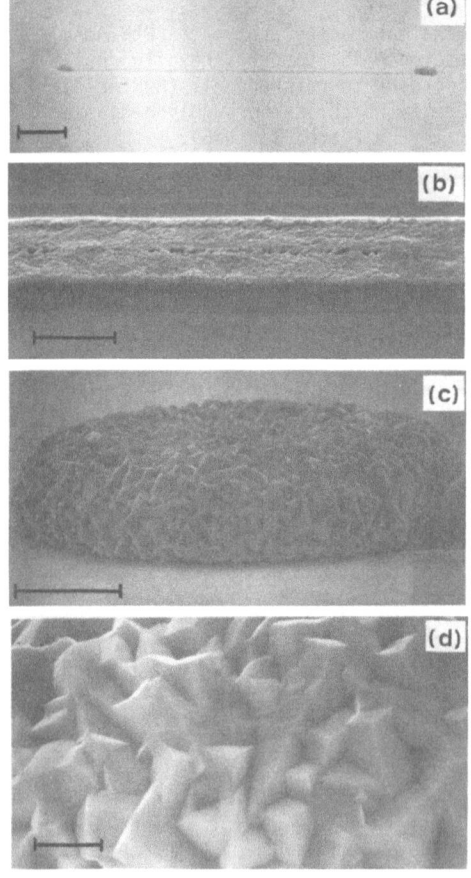

Fig. 2 (a) Cu ohmic contact by pyrolytic LCVD of CuHFAcAc on glass. Writing speed 17 microns per second, power density of 514.5 nm light is 60 kW cm^{-2}. The cell temperature is 90 ^{o}C. The marker indicates a length of 100 microns on the scanning electron microscope (SEM)picture. (b) An enlargement of the middle section of Fig. 2a showing the regularity of the deposit. The marker represents 10 microns. (c) Enlargement of the starting point of the Cu contact shown in Fig.2a. The marker indicates again a 10 micron distance. (d) An enlargement of Fig. 2c showing the crystalline nature of the pyrolytically deposited copper. The marker now represents 1 micron.

be seen from the enlarged SEM picture (Figs. 2C and D). We have previously found in the LCVD of Sn and Pt (14) that whereas photolytic LCVD often gives rise to amorphous deposits, pyrolytic LCVD gives rise to polycrystallin deposits. One possible explanation for this might be that only on a heated surface will the deposited metals have enough energy to rearrange to the crystallin structure. The energy necessary for this rearrangement varies from metal to metal. Several other phenomena are observed when making such deposits as shown in Fig. 2. The effect of adding 1 atmosphere of an inert gas generally results in a better defined deposit. This may in part be due to the reduction of the mean free path. Furthermore,induction times are observed in this pyrolytic LCVD. These times can vary from apparently instantaneous deposition to induction periods of several minutes, and generally decrease with increasing power density. The sudden initiation of the pyrolytic LCVD is clearly indicated by a large increase of scattered light from the deposited metal clusters. Such inhibition periods are expected to depend strongly on impurities present as well as the absorption properties of the metal-organic compound absorbed on the surface which itself varies with temperature and pressure. For every writing speed there appears to be an optimal light intensity at which the best deposits occur. For instance under the conditions of Fig. 2 at light intensities of 100 kW cm^{-2} or more the deposits are more granular and less regular overall. We have succeeded in making deposits at speeds up to 0.5 mm sec^{-1}. The deposits must stick to the surface and are exposed to a sticky tape tests. The sticky tape does not enable us to lift the deposit off the substrate.

One of the most important bulk characteristics of these metallic deposits is their bulk resistance. In the pyrolytic LCVD of copper we obtain resistances of the order of 20 times bulk, a value which we expect to be able to improve upon significantly in the near future. Besides the microstructure of the deposit, this electrical resistance will be influenced significantly by the amount of impurities incorporated in the copper. Decomposition of this particular metal-organic compound can give rise to F, O and C in the copper. We are planning to analyse these impurities by Auger spectroscopy,and try to correlate them with the bulk resistivity. Generally pyrolytic LCVD leads to less impurities in the deposit than photolytic LCVD, probably due to the fact that the higher temperatures allow the impurities which are more volatile than the metallic deposit to escape from the surface. Thus,in future LCVD one might search for an optimum in surface temperature while performing photolytic LCVD by using two wavelengths, one to dissociate photolytically the metal organic, and the other to heat the surface to a point where less impurities are incorporated in the deposit. If these temperatures are less than the temperatures needed for efficient pyrolysis the substrate will then also be less distorted by the high temperatures needed for the latter.

Photolytic LCVD of CuHFAcAc was also attempted.at power densities around 600 W cm^{-2} and writing speeds of 0.3 micron per second. The bulk resistance of these deposits was high and they contained large quantities of impurities.

4. Platinum deposition

An extensive discussion of platinum pyrolytic and photolytic discussion and the bulk resistivities obtained will appear in the near future.(18). Thermal deposition in preliminary measurements gave resistivities of less than 50 times bulk. Osgood, Gilgen and coworkers (9) have obtained resistivities of about 2 times bulk for conditions at which both pyrolysis and photolysis may play a role at 350 nm.

5. Grating made by LCVD

Fig. 3 shows a grating of Ga metal deposited from trimethylgallium at 10 torr by photolytic LCVD at 257 nm. Interference between the laser beam and its reflection gives rise to such gratings which we have produced with a line width of less than 0.2 microns. The power was 100 W cm^{-2}.

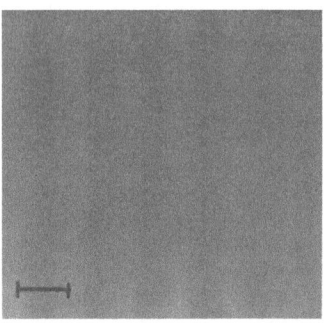

Fig. 3. Grating of Ga made by photolytic LCVD from trimethylgallium at 257 nm. Organometallic vapor pressure is 10 torr. Power density 100 W cm^{-2}. The marker indicates a length of 0.2 microns.

4. Tin deposition

In Fig. 4 we show an interesting detail which is indicative for thin film growth in the pyrolytic LCVD of tin. The greyish zone on the right side of the picture is a thin layer of Sn. As the polycrystalline Sn particles move about the surface (Brownian motion) the thin layer of Sn sticks to these particles (the black dots) which then grow, leaving wider and wider trails of surface denuded of tin in the process.

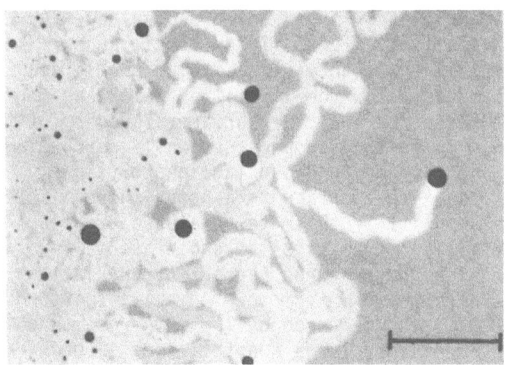

Fig. 4. Transmission electron micrograph showing the growth of Sn particles at the fringe of a pyrolytic LCVD deposit. Laser power density 50 kW cm^{-2} and irradiation time 3 s.

5. References

1. D.J. Ehrlich, R.M. Osgood and T.F. Deutsch, J. Vac. Sci. Technol. 21, 23(1982).
2. R.M. Osgood, Ann. Rev. Phys. Chem. 34, 77(1983).
3. Laser Diagnostics and Photochemical Processing for Semiconductor Devices, R.M. Osgood et. al. Eds. (North Holland, New York, 1983).
4. Laser Controlled Chemical Processing of Surfaces, A.W. Johnson et. al. Eds. (North Holland, New York, 1984).

5. Laser Processing and Diagnostics, D. Bäuerle Ed., Springer Ser. Chem. Phys. 39, Springer Verlag, Berlin (1984).
6. D.J. Ehrlich and J.Y. Tsao, "Laser Direct Writing for VLSI", in VLSI Electronics: Microstructure Science, Vol 7, N.G. Einspruch Ed. (Academic Press, New York, 1983)
7. Qiu Mingxin, R. Monot and H. van den Bergh, Scientia Sinica A, 27, 531(1984).
8. J.-M. Philippoz, Diplôme, EPFL (1982).
9. H. Gilgen and R.M. Osgood personal communication.
10. F.A. Houle et. al. Appl. Phys. Lett. in press.
11. D. Bäuerle, personal communication.
12. D.J. Ehrlich, personal communication.
13. M.R. Aylett and J. Haigh, in ref. 5 pp. 263-268.
14. D. Braichotte and H. van den Bergh, in ref. 5 pp. 183-187.
15. H.H. Gilgen, C.J. Chen, R. Krchnavek and R.M. Osgood, in ref. 5, pp. 225-233.
16. S. Okeya and S. Kawaguchi, Inorg. Synth. 20, 65(1980).
17. R.L. Belford, A.E. Martell and M. Calvin, J. Inorg. and Nucl. Chem., 2, 11(1956).
18. D. Braichotte and H. van den Bergh, to be published.

Electro-Optically Induced Optical Waveguide in KNbO₃

J.-C. Baumert[1,2], C. Walther[1], P. Buchmann[2], H. Melchior[2], and P. Günter[1]

[1]Laboratory of Solid State Physics, [2]Institute of Applied Physics, Swiss Federal Institute of Technology, CH-8093 Zürich, Switzerland

Electro-optically induced optical waveguiding in the wavelength range 0.5 μm - 1.3 μm and cut-off modulation in a structure based on KNbO₃ (electro-optic coefficient r_{33} = 64 pm/V) single crystals are reported.

Potassium niobate (KNbO₃, point group symmetry mm2 at room temperature) is a very interesting electro-optical material for both bulk and waveguide applications, because of its large electro-optic [1] and nonlinear optic [2,3] coefficients, good photorefractive properties [4,5,6] and high damage threshold (60 MW/cm² pulsed at λ = 0.86 μm). These properties make KNbO₃ attractive for thin film waveguides, such as electro-optic modulators, which would benefit from high figures of merit $n_3^3 r_{33}$ = 680 pm/V and $n_4^3 r_{42}$ = 4350 pm/V (n_3 = 2.1683 is a principal refractive index, n_4 = 2.254 is an average refractive index in the bc-plane and r_{42} = 380 pm/V [1]) compared to $n_3^3 r_{33}$ = 341 pm/V [7] for LiNbO₃, or an efficient frequency doubler for Al$_x$Ga$_{1-x}$As-semiconductor lasers, allowing collinear phase-matched type I interaction around room temperature in this wavelength range. TUCKER et al. [8] (1974) observed optical waveguiding in naturally formed planar sheet domains in KNbO₃.

In this paper we report on the first waveguides in KNbO₃ induced by the electro-optic effect. In order to use the electro-optical coefficient r_{33} in KNbO₃, a crystal plate was cut normal to the b-axis, and two electrodes with a width of (s-h) = 100 μm, separated by a gap of width 2h = 10 μm, were deposited on the polished b-face (see Fig. 1). The edges of the electrodes are parallel to the a-axis. The horizontal (parallel to the c-axis) component $E_x(x,y)$ of the applied electric field yields an increase of the refractive index n_c of the crystal in the gap region given by

$$\Delta n_c(x,y) = \frac{1}{2} n_c^3 r_{33} E_x(x,y) = A \cdot E_x(x,y) \tag{1}$$

with A = 3.262·10⁻⁴ μm/V for λ = 0.633 μm. The refractive index change Δn_b, due to the vertical electric field $E_y(x,y)$ has been neglected because of the small electro-optic coefficient r_{23} = 1.3 pm/V [1]. Therefore with this type of waveguide only TE-modes propagating along the a-axis are guided.

A preferentially single domain KNbO₃ crystal was grown by a top seeded high-temperature melt pulling technique [9]. Chips with a size of 4x3.4x0.7 mm were cut from the crystal and orientated by X-ray and preferential etching methods [10] . After surface polishing the remaining domains were removed in a strong polarizing dc field near the Curie temperature. Opposite ends of the single crystals were polished in order to allow for end-fire coupling of laser light. This process, however, has caused stress induced microdomains along the edges of the facet. For electrode preparation, a thin film of Chrom/ Gold was deposited by electron beam evaporation on the b-cut surface. MP 1350J positive resist was applied and baked very carefully to prevent creation of new domains (heating/cooling cycle with dT/dt <2°C/min). The electrode struc- ture was patterned and the metal film etched. The electrodes had a length of 3 mm (Fig. 1). The samples were mounted on a ceramic substrate and contacted using copper wire and silver paste.

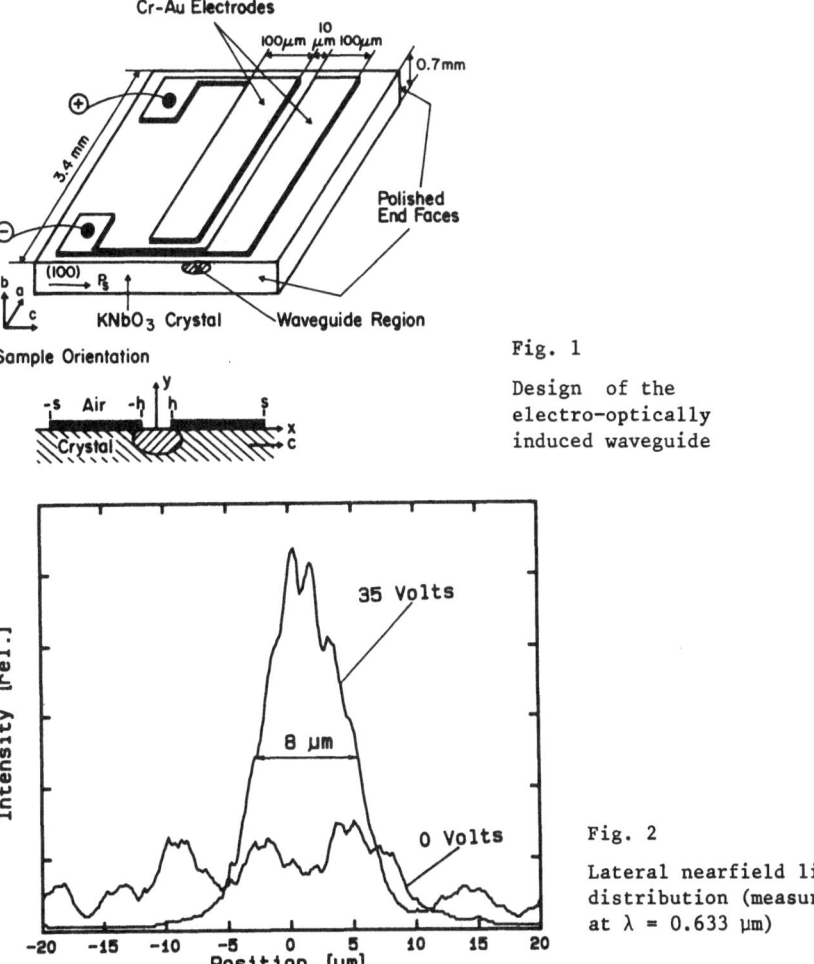

Fig. 1

Design of the electro-optically induced waveguide

Fig. 2

Lateral nearfield light distribution (measured at λ = 0.633 µm)

TE-polarized light from different laser sources (λ = 1.3 µm, 1.064 µm, 0.86 µm, 0.633 µm, 0.532 µm) was coupled into the electro-optically induced waveguide by end-fire coupling. The outcoupled beam was monitored with a vi- dicon and a linear photo-diode array. With no electric field applied, we could observe only some lightspots, caused by diffraction at stress-induced

45

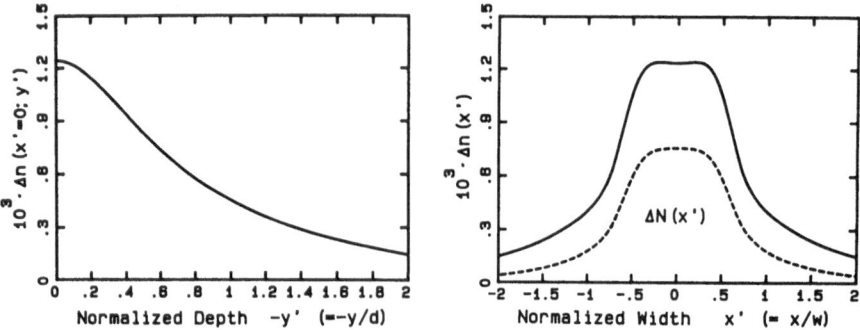

Fig. 3   Refractive index depth (y') and lateral (x') profiles for an applied voltage of 35 Volts and λ=0.633 μm (calculated). $\Delta N(x')$ (dotted line)
is the effective lateral index.

domains at the crystal endfaces. Increasing the applied voltage up to 30
volts, an on-off ratio of 12 dB could be measured, clearly demonstrating a
field-induced increase of the refractive index n_c between the two electrodes.
Figure 2 shows the measured lateral nearfield light distribution at 0.633 μm
for an applied voltage of 0 and 35 volts. This distribution was also calculated. The refractive index distribution was evaluated using eq. (1) and
conformal mapping techniques [11,12] in order to calculate the electric field
component $E_x(x,y)$ inside the crystal. The electric field-induced refractive
index changes Δn are shown in Fig. 3 for the x and y directions. The coordinates have been normalized to the effective depth (d = 12,5 μm) and width
(w = 11,2 μm).

The effective index approach [13] together with a new method [14] for solving the scalar wave equations were used to calculate the intensity distribution of the fundamental guided mode (TE_{oo}). These results are shown in Fig.
4 and are in excellent agreement with the measurements.

The properties of this waveguide as a cut-off modulator have also been
studied. Therefore, the outcoupled light was imaged onto a photomultiplier.
Because the fundamental mode is guided, for applied voltages above 10 Volts

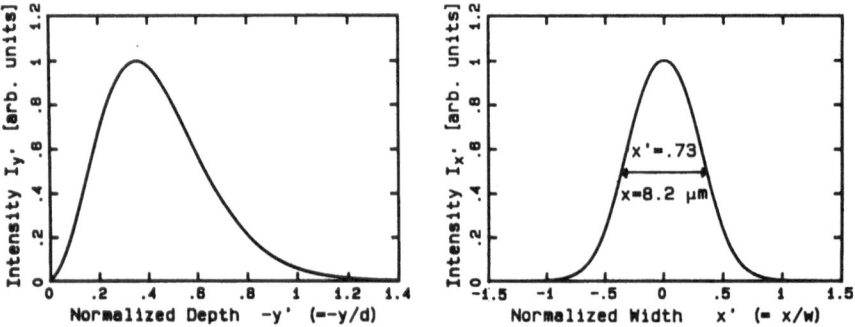

Fig. 4   Intensity distribution I_x', I_y' of the fundamental guided mode
         (calculated)

only, we have biased the waveguide with 10 Volts DC. The modulation voltage was either a sinusoidal or a square wave with an amplitude of 20 Volts. The bandwidth was measured to be 1.4 MHz corresponding to a rise time of 700 ns (see Fig. 5). This frequency was limited by the rather high distributed capacity (C = 315 pF), caused by the high effective dielectric constant ($\varepsilon_0(1 + \sqrt{\varepsilon x \varepsilon y'}) = 2.1$ nF/m [15]) and the fact that four such electrode structures were connected electrically parallel on the same crystal. By using a single structure on an a-cut crystal and reducing the electrode width to 10 µm a bandwidth of 500 MHz could be reached.

In conclusion, technologically controlled optical waveguiding has been reported in $KNbO_3$, for the first time to our knowledge. Thanks to the very large electro-optic coefficients of $KNbO_3$, the induced light-guiding can be achieved at relatively low drive voltages. Since the device can also be used as a cut-off modulator, it may be interesting as an integrated optical gate or switch. The waveguide technique could also be very useful in other devices making use of the large electro-optic and nonlinear optic coefficients and the favorable photorefractive properties.

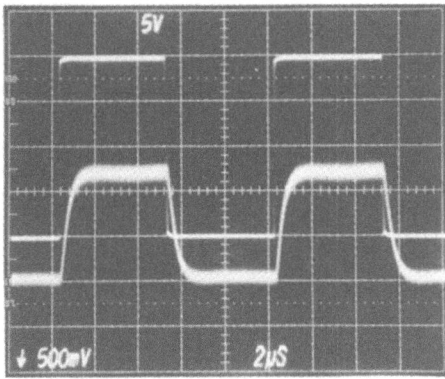

Fig. 5

Modulator response (measured)
Upper trace: Modulation signal
Lower trace: Detected light
                      signal

Acknowledgements

The authors are very grateful to H. Arend and H. Wüest for the crystal growth and to M. Bachmann for the technical assistance. This work was supported by the NSF Research Program: "Microelectronics and Optoelectronics" (NFP 13). P. Buchmann acknowledges support by the Swiss PTT.

References

1. P. Günter: Electro-Optics/Laser International 76 UK, 121 (1976)
2. Y. Uematsu: Jap. J. Appl. Phys. 13, 1362 (1974)
3. J.-C. Baumert, J. Hoffnagle and P. Günter: Proceedings of ECOOSA'84 conference, Amsterdam, SPIE Vol. 492 (to be published)
4. P. Günter: Physics Reports, 93, 199 (1982)
5. P. Günter and F. Micheron: Ferroelectrics, 18, 27 (1978)
6. P. Günter: Optics Letters 7, 10 (1982)
7. Landolt-Börnstein: New Series, Vol. 11 (Springer-Verlag, Berlin, 1979)
8. J.R. Tucker, A.B. Chase and S.R. King: Digest of Techn. Papers MB12-1 (1974)

9. U. Flückiger and H. Arend: J. of Cryst. Growth 43, 406 (1978)
10. E. Wiesendanger: Czech. J. Phys. B, 23, 91 (1973)
11. J.S. Wei: IEEE J. Quant. Electron. QE-13, 152 (1977)
12. O.G. Ramer: IEEE J. Quant. Electron. QE-18, 386 (1982)
13. see e.g. M.J. Adams: An Introduction to Optical Waveguides (John Wiley &
        Sons, Chichester 1981)
14. J.-C. Baumert, J. Hoffnagle and C. Walther (to be published)
15. E. Wiesendanger: Ferroelectrics 6, 263 (1974)

Single-Mode Magneto-Optic Waveguides in Multiple-Layer Garnet

E. Pross, H. Dammann, W. Tolksdorf, M. Zinke

Philips GmbH, Forschungslaboratorium Hamburg, Vogt-Kölln-Straße 30
D-2000 Hamburg 54, Fed. Rep. of Germany

A fabrication procedure for growing Yttrium Iron Garnet (YIG) single-mode waveguides compatible with single-mode fibers has been developed. The required $\Delta n \sim 4 \cdot 10^{-3}$ is controlled by growing double YIG layers with slightly different composition from the same melt on Gadolinium Gallium Garnet (GGG) substrates. Single-mode operation is confirmed experimentally.

1. INTRODUCTION

Magneto-optic film waveguides are the basic elements for the realization of planar, nonreciprocal components such as waveguide isolators [1] and circulators which are expected to play a key role in advanced fiber communication and fiber sensor systems. The operation of these components is based on the Faraday effect in magnetic crystals such as pure or substituted Yttrium Iron Garnet (YIG) [2] which is transparent in the wavelength range 1.1-5 μm.

In communication systems, the magneto-optic waveguides have to be combined ('integrated') with other elements as, e.g., single-mode fibers. In order to facilitate simple and efficient 'end-butt' coupling, the waveguide structures of the different components should be similar. The core diameters of single-mode fibers are in the range of 3-10 μm, and the refractive index steps Δn are on the order of $\Delta n \approx 4 \cdot 10^{-3}$. Corresponding values are desired for the magneto-optic waveguides. This paper describes a new fabrication technique for such single-mode waveguides in YIG and the experimental demonstration of their operation.

2. DESIGN OF SINGLE-MODE WAVEGUIDES IN YIG

YIG planar film waveguides (Fig. 1a) are grown on Gadolinium Gallium Garnet (GGG) substrates by liquid phase epitaxy (LPE) [3]. Due to the large refractive index difference between the YIG film and the GGG substrate ($\Delta n \approx 0.28$), such a waveguide with a thickness of several micrometers ($d_2 \sim 3$-10 μm) supports many modes. In order to achieve single-mode operation in a waveguide of that thickness, Δn has to be reduced to $\Delta n \sim 4 \cdot 10^{-3}$. This can be realized by introducing an intermediate YIG layer with a slightly decreased refractive index as shown in Fig. 1b. Then, the upper YIG layer (called YIG 2) is a single-mode waveguide with the additional YIG layer (called YIG 3) as a new substrate. The thickness d_3 of YIG 3 is not critical but should be larger than 3 μm because the mode penetrates considerably into that layer.

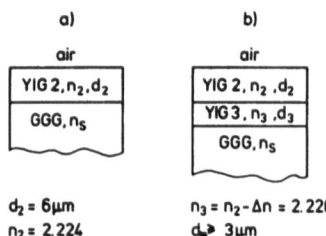

a)

air

| YIG 2, n_2, d_2 |
| GGG, n_s |

$d_2 = 6\mu m$
$n_2 = 2.224$
$n_s = 1.945$

b)

air

| YIG 2, n_2, d_2 |
| YIG 3, n_3, d_3 |
| GGG, n_s |

$n_3 = n_2 - \Delta n = 2.220$
$d_3 = 3\mu m$

Fig. 1
Planar waveguide structures in YIG
a) YIG mono-layer on GGG, forming a multimode waveguide
b) Double YIG layer on GGG. The upper layer
   YIG 2 is a single-mode waveguide with
   YIG 3 as the new substrate. Values given
   for refractive index n correspond to
   $\lambda = 1.15$ μm.

3. SAMPLE PREPARATION

YIG waveguiding layers were grown by LPE according to [3] from
a mixture of lead oxide/boron oxide as solvent. Substrates
32 mm in diameter were horizontally dipped into the melt and
rotated during growth with an alternating direction of rotation. The refractive index of the layers depends on the incorporation of substituents such as Bismuth (Bi), Gallium (Ga)
or Lead (Pb). We found that the refractive index is adjustable
within a range of about $\delta n \gtrsim 10^{-2}$ simply by choosing an appropriate lead incorporation. This is realized by a proper set-
ting of one or both of the following two LPE parameters:

1. The supercooling of the melt.
2. The rotation rate of the sample.

Increasing of either parameter value results in increased incorporation of Pb. Each parameter may be varied independently
of the other.

We grew modified YIG mono-layers of composition $Y_{3-x}Pb_xFe_5O_{12}$
with lead content x in the range $0.006 < x < 0.106$. The lead
content was measured by means of Electron Probe Microanalysis
(EPMA) [4]. Refractive indices of the samples were measured by
dark m-line-spectroscopy [5] at $\lambda = 1.15$ μm with an accuracy
of better than $1 \cdot 10^{-3}$. The results are presented in Fig. 2,
showing a fairly linear dependence of the refractive index n
on lead content x. The samples investigated were grown from
two different melts; the degree of Pb incorporation depends
not only on growth conditions but on melt composition as well.

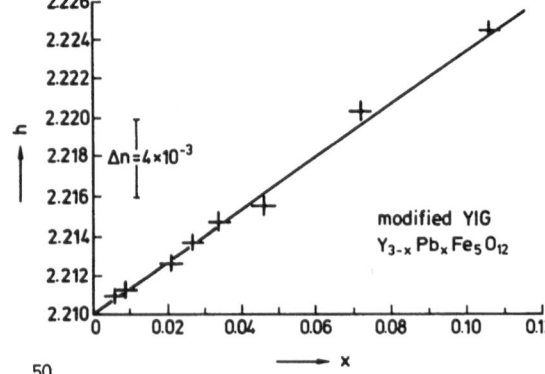

modified YIG
$Y_{3-x}Pb_xFe_5O_{12}$

$\Delta n = 4 \times 10^{-3}$

Fig. 2
Dependence of refractive index n on lead content x for
epitaxially grown YIG layers
$(\lambda = 1.15$ μm).

For samples grown from the same melt, refractive indices ranging from 2.211 to 2.221 ($\delta n = 1 \cdot 10^{-2}$) were obtained by varying
the supercooling of the melt ΔT_s from 3° to 57° while the
rotation rate (ω) was kept constant at 60 min^{-1}. Next, supercooling was kept constant at 57° while ω was changed from
30 min^{-1} to 160 min^{-1} yielding an increase of the refractive
index from 2.220 to 2.224 ($\delta n = 4 \cdot 10^{-3}$).

The results shown in Fig. 2 correspond to mono-layers of modified YIG on GGG as shown in Fig. 1a. Double-layers as sketched
in Fig. 1b have been realized using the same technique, but
stepwise changing the LPE conditions during the growth of one
and the same sample according to the values obtained by mono-
layers.

An alternative method for the manufacturing of modified YIG
double-layers is to fabricate the layers from two melts with
different compositions [6]. Refractive index differences Δn
can be larger using different melts, but an accuracy of $1 \cdot 10^{-3}$
is hardly achievable by this method.

4. WAVEGUIDING EXPERIMENTS

Single-mode operation of modified YIG layers was experimentally tested. Fig. 3 shows a photograph of the angular mode
spectrum of a single-mode waveguide as displayed by m-line
spectroscopy ($\lambda = 1.15$ µm). The sample under investigation
had the following parameters: Thickness of waveguiding layer
$d_2 = 3.5$ µm, $d_3 = 3.0$ µm, and refractive index step $\Delta n = 4 \cdot 10^{-3}$
between the two YIG layers. The only dark m-line visible cor-
responds to the single mode propagating in the upper (YIG 2)
layer.

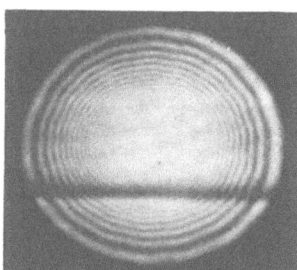

Fig. 3
M-line spectrum of a single-mode waveguide in
YIG. The dark line corresponds to the mode
propagating in the upper (YIG 2) layer.

In a second experiment, the light from a 1.3 µm-LD was trans-
fered to the waveguides by end-fire coupling, and images of
the end-faces of the waveguides were projected onto the target
of an IR-TV-camera. Fig. 4 shows a photograph taken from the
TV-screen. The waveguide is equivalent to the one used in the
m-line experiment, except that waveguide thickness $d_2 = 7.0$ µm
here and total layer thickness is 14 µm. Clearly, light is
only guided by the upper layer.

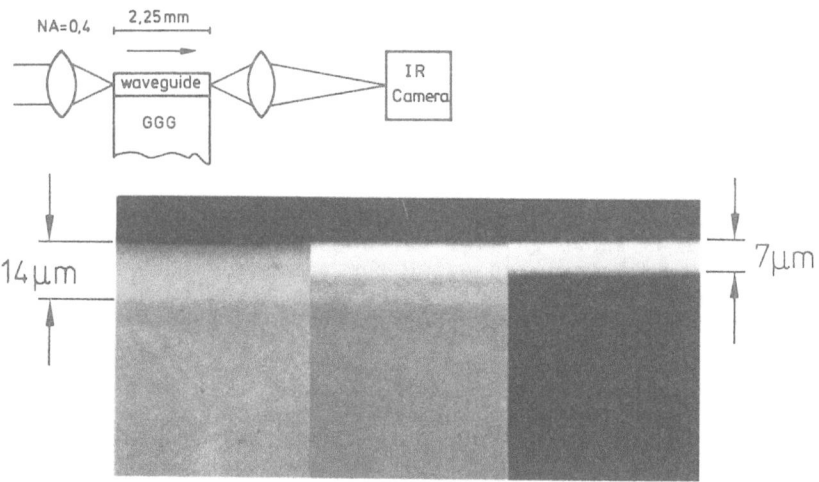

Fig. 4
End-fire coupling to a single-mode YIG waveguide, output-face is shown.
Left: Sample illuminated without coupling. The YIG layers appear brighter
than the GGG substrate.
Center: Coupling established. The bright streak about 7 μm thick indi-
cates that light is only guided by the upper layer.
Right: Sample illumination switched off.

5. CONCLUSION

Single-mode operation in magneto-optic waveguides suitable for
direct coupling with single-mode fibers has been demonstrated
experimentally. These waveguides can now be used as basic
elements for waveguide isolators.

6. ACKNOWLEDGEMENT

The authors thank Mrs. I. Bartels for carefully preparing the
samples, P. Willich for performing the EPMA analysis and
C.P. Klages for helpful discussions.

The described work was sponsored by the German Federal Ministry of Research and Technology (BMFT) under grant number
TK 0251 5. Only the authors are responsible for the contents
of this publication.

REFERENCES

[1] See e.g. S.T. Kirsch et al.: J. Appl. Phys. 52, (1981) 3190.
[2] G. Winkler: Magnetic Garnets, Vieweg Verlag 1981.
[3] P. Klages, W. Tolksdorf, G. Kunat, J. of Crystal Growth
    65, (1983) 556.
[4] P. Willich, W. Tolksdorf, D. Obertop, J. of Crystal Growth
    53, (1981) 483.
[5] R. Ulrich and R. Torge, Appl. Optics, 12, (1973) 2901.
[6] A. Shibukawa and M. Kobayaski, Appl. Optics 20, (1981) 2444.

Normalized Diagrams for Diffused Waveguides Optical Properties: Application to Ti:LiNbO₃ Electrooptic Directional Coupler Design

L. Riviere, A. Carenco, A. Yi-Yan, and R. Guglielmi

Centre National d'Etudes des Télécommunications, Laboratoire de Bagneux
196 rue de Paris, F-92220 Bagneux, France

Optical parameters of diffused waveguides can be simply predicted by normalized diagrams, directly related to fabrication conditions. The derived characteristics of electrooptic directional couplers are in good agreement with experience.

I-Introduction

Ti-indiffused LiNbO₃ waveguides have largely proved their worth in integrated optics. An optimized design of devices requires a good knowledge of the relations between the fabrication parameters and the optical properties. Here is presented a simple universal tool for rapid estimation of devices characteristics, based on normalized diagrams directly related to diffusion conditions. The method is applied to a LiNbO₃ electrooptic directional coupler (EDC). The diagrams provide the mode size (Γ_0) of a diffused waveguide and the coupling length (ℓ_c) of an EDC, as a function of fabrication parameters. From these optical characteristics are derived the best conditions for insertion loss and driving voltage minimization of a fiber-pigtailed active device.

II-Mode Size of a Diffused Waveguide

The index profile of a diffused waveguide (Fig. 1) can be written [1]:

$$n^2(y,z)=n_b^2 + 2\,n_b\,C\,\frac{\tau}{D_z}\,\exp\left[-\left(\frac{z}{D_z}\right)^2\right]\,\frac{1}{2}\left\{\text{erf}\left[\frac{1}{D_y}(y+\frac{W}{2})\right]-\text{erf}\left[\frac{1}{D_y}(y-\frac{W}{2})\right]\right\}$$

where n_b is the substrate index, D_y and D_z the diffusion lengths (depending on time and temperature), τ and W the thickness and the width of the Ti stripe, C a typical value characterizing the linear relationship between the extraordinary index difference and Ti-concentration. (C is a function of wavelength [2] and of Ti density in the deposited layer).

In order to reduce the number of parameters, two more approximations are made :

1°) The evanescent field in the superstrate (SiO₂ or air) is neglected. This can be justified by the small index difference in a diffused waveguide as compared to that between the substrate and the superstrate.

2°) The D_z/D_y ratio is taken as a constant (=1.2) as suggested by the experimental results [2] obtained in the practical range of diffusion temperature (1000°C to 1050°C).

Using these approximations, the solution of the scalar wave equation for a weakly guided mode only depends on two normalized parameters :

$$\tau_n=(2\pi/\lambda)^2\,2\,n_b\,C\,D_z\,\tau \qquad \text{normalized thickness}$$

$$W_n= W/D_z \qquad \text{normalized width}$$

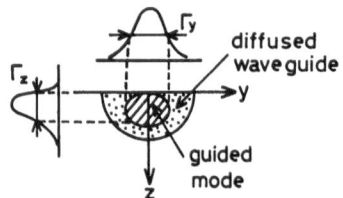

Fig. 1 : Definition of mode size

The mode size $\Gamma_0 = \sqrt{\Gamma_y \Gamma_z}$, where Γ_y and Γ_z are the full width at half maximum intensity in the y and z directions respectively (Fig. 1), is normalized :

$$\Gamma_{on} = \Gamma_0 / D_z$$

Γ_{on} is calculated using the effective index method [3]. Analytical expressions have been developed to allow a simple and accurate determination of effective indices of the multiple gaussian planar waveguides [4]. The lateral variation of the guided field is finally obtained from the effective index profile with help of the multilayer technique [5]. Curves of constant Γ_{on} are plotted in Fig. 2 as a function of τ_n and W_n.

In this normalized diagram, the single mode region is delimited by two broken lines (-·-), calculated by an arbitrarily chosen physical criterion ($\Delta N_{eff} > 0.05 \, \Delta n(0,0)$). It is noteworthy that both limits are well approximated by two hyperboles : $W_n \cdot \tau_n = 14$ and 38 for the first and second mode respectively (dashed line : ---). It turns out, from these very simple expressions, that the number of guided modes
- does not depend on time and temperature of diffusion
- directly depends on the section (W_τ) of the initial Ti stripe.
This rough estimation is only valid when $W_n < 3$ (i.e. when lateral diffusion cannot be neglected).

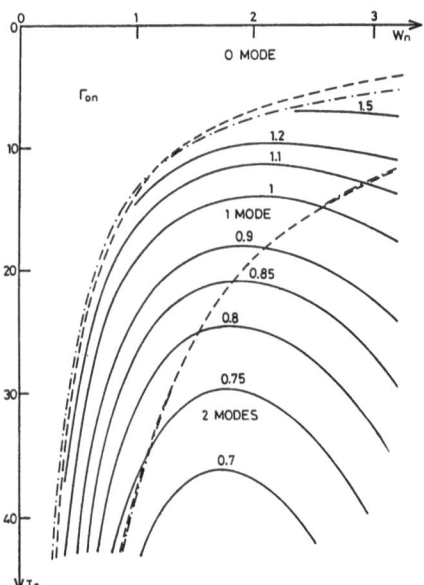

Fig. 2 : Normalized mode size Γ_{on}

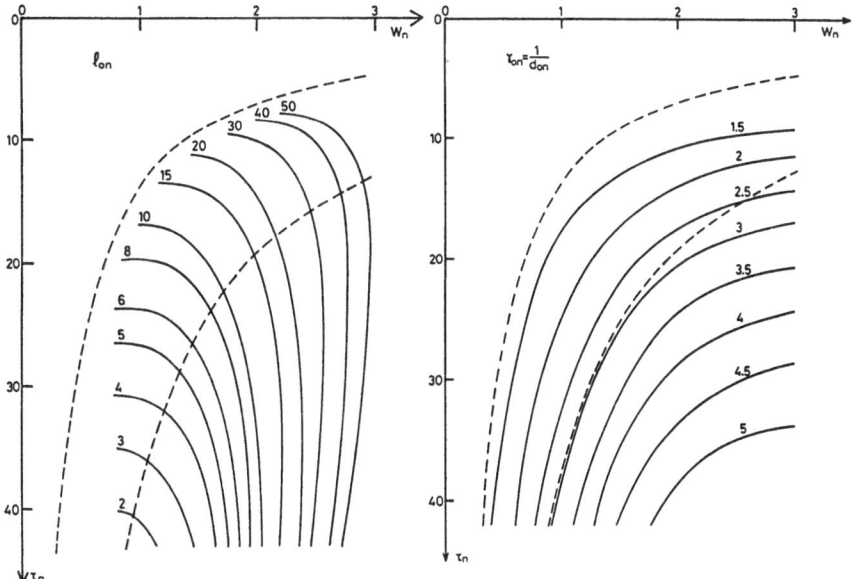

Fig. 3 : Normalized diagrams for evaluation of the coupling length

III-EDC Coupling Length

In an EDC, with a gap d between Ti stripes, the index profile results from
the linear superposition of the two diffused waveguides. With the well-es-
tablished assumption on coupling length :

$$\ell_c = \ell_0 \exp(d/d_0)$$

it is possible to describe ℓ_c evolution using only two parameters (ℓ_0 and d_0).
Normalized expressions for ℓ_0, ℓ_c, d_0 and d are :

$$\ell_{on} = \frac{\lambda}{n_b D_z 2}\, \ell_0 \; ; \; \ell_{cn} = \frac{\lambda}{n_b D_z 2}\, \ell_c \; ; \; d_{on} = \frac{d_0}{D_z} \; ; \; d_n = \frac{d}{D_z}$$

The curves of constant ℓ_{on} and $\gamma_{on} = 1/d_{on}$ are plotted in Fig. 3 as a
function of τ_n and W_n. They are obtained by adjusting the normalized exponential relationship to 4 values of ℓ_{cn} computed for 4 different values of d_n
(0.5 ; 0.75 ; 1 ; 1.25). Any extrapolation of d_n out of the (0.3-1.5) range
could lead to unacceptable errors.

IV-Applications

IV-a-Total Insertion Loss

Three main mechanisms contribute to total loss, namely reflection, propagation and coupling to a fiber. While reflection loss has a constant value of
about 0.16dB/face, the two other factors are related to Γ_0. For this purpose,
evolution of the mode size as a function of the stripe geometry appears in
Fig. 4 for two different diffusion conditions (1000°C and 1050°C, 16 hours).
The curves are derived from the normalized diagrams using the diffusion parameters measured in the laboratory (TM mode, λ=1.55µm ; n_b=2.1383 ; C=0.77 ;
E_{az}=2.76eV ; $\mathfrak{D}_{oz}$=2.3 10^{10}µm²/h). Low coupling losses are obtained when the
mean mode size in the diffused waveguide is matched to that of the fiber
($\sim$ 5µm). Low propagation losses need a well-confined mode close to the second
mode cut-off value (region arbitrarily delimited by $34 \leq W_n \cdot \tau_n \leq 42$). As illustra-

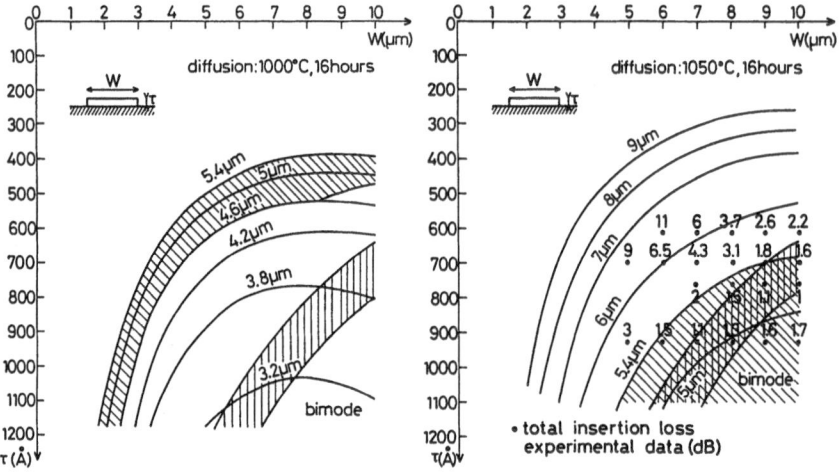

Fig. 4 : Mode Size of Diffused Waveguides : ▧ Low coupling loss region
                                              ⬚ Low propagation loss region

ted in Fig. 4, a deep diffusion leads to an overlap of the low coupling loss
region with the region of low propagation loss. These predictions have been
corroborated with practical devices yielding a fiber-guide-fiber total insertion loss around 1dB as illustrated by the experimental data superimposed in
Fig. 4b.

IV-b-EDC Driving Voltage

The driving voltage ΔV of an EDC depends both on the electric field induced
by electrodes and on the optical guided-mode properties. The electrode geometry generally results from bandwidth considerations. The waveguide gap d
is adjusted to obtain the optimum value of ℓ_c, minimizing ΔV, according to
the electrodes configuration (uniform or alternated Δβ). The normalized diagrams of Fig. 3 allow this determination as a function of diffusion conditions. Knowing the mode position, mode size and electrode geometry, ΔV can

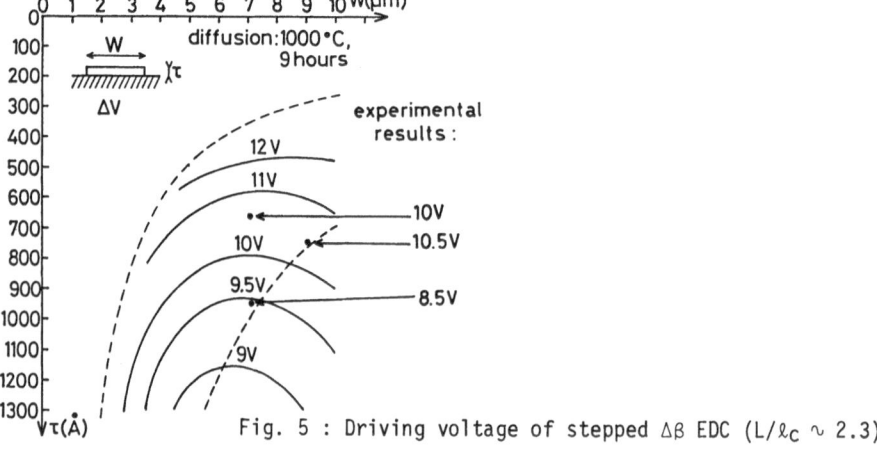

Fig. 5 : Driving voltage of stepped Δβ EDC (L/ℓ_c ∿ 2.3)

be computed [6,7]. Because ΔV does not depend critically on modal ellipticity (Γ_y/Γ_z), this parameter can be set to a mean value ($\sim$ 1.6). Computed results appear in Fig. 5 for a 1000°C, 9 h diffusion and symmetrical electrodes (length 9mm, width 30μm, gap 3μm). More generally, in the range of useful diffusion conditions, an interesting correlation has been found to occur between Γ_0 and ΔV, once the electrode geometry is fixed. For example, with the previously described electrodes,

$$\Delta V(\text{volts}) = 2.48 \ \Gamma_0 + 2.1 \qquad (\Gamma_0 \text{ in } \mu\text{m})$$

yields a fairly good approximation ($\pm$ 0.5V) of the theoretical evolution of ΔV versus the diffusion conditions. This simple expression points out the role of mode confinement on driving voltage.

V-Conclusion

In conclusion, it has been demonstrated that the optical parameters of diffused waveguides (mode size, coupling length) can be simply predicted by normalized diagrams. Application to the design of an electrooptic directional coupler (insertion loss, driving voltage) illustrates the interest of the method. Experimental results are in good agreement with theoretical estimations.

References

[1] Burns W.K., Klein P.H. and West E.J. : "Ti Diffusion in LiNbO$_3$:Ti Planar and Channel Optical Waveguides", in J. Appl. Phys., Vol. 50, p 6175, 1979.
[2] Fouchet S., Guglielmi R., Yi-Yan A. and Carenco A. : "Wavelength dispersion of Ti:LiNbO$_3$ Waveguides", in 2nd European Conf. on Integrated Optics, Firenze, IEE Conf. Publication n° 227, p 50, 1983.
[3] Hocker L. and Burns.W.K.: "Mode Dispersion in Diffused Channel Waveguides by the Effective Index Method", in Appl. Opt., Vol. 16, p 113, 1981.
[4] Riviere L, Yi-Yan A and Carru H. : "Properties of Single Mode Optical Planar Waveguides with Gaussian Index Profile", to be published in IEEE J. of Lightwave technology.
[5] Vassel M.O. : "Structure of Optical Guided Modes in Planar Multilayers of Optically Anisotropic Materials" in J. Opt. Soc. Am., Vol. 64, p 166, 1974.
[6] Ramer O.G. : "Integrated Optic Electrooptic Modulator Electrode Analysis", in IEEE J. Quant. Electron., Vol. QE-18, p 386, 1982.
[7] Sabatier C. and Caquot E. : "Influence of a Dielectric Buffer Layer on the Field Distribution in an Electrooptic Guided-wave Device", submitted to publication.

Formation and Analysis of Tapers in Proton-Exchanged Lithium Niobate Waveguides

G. Stewart and A.C.G. Nutt

Department of Electronics and Electrical Engineering, University of Glasgow
Glasgow, G12 8QQ, United Kingdom

Tapered waveguides for directional couplers have been formed in $LiNbO_3$ by proton exchange and analysed by observing interference effects in light reflected from the guide.

1. Introduction

Recently there has been much interest in the stuay of the proton exchange (PE) method for waveguide formation in lithium niobate substrates[1]. Such waveguides have approximately a step index distribution with index increments of >0.1 (for extraordinary index). Here we report the formation of tapered regions in the waveguide which have application in the construction of taper-velocity couplers.

2. Formation and analysis of tapers

Tapered waveguiding regions such as illustrated in Fig. 1 (inset) may simply be formed by partially immersing the $LiNbO_3$ sample in the Benzoic acid, the taper being formed at the meniscus of the acid.

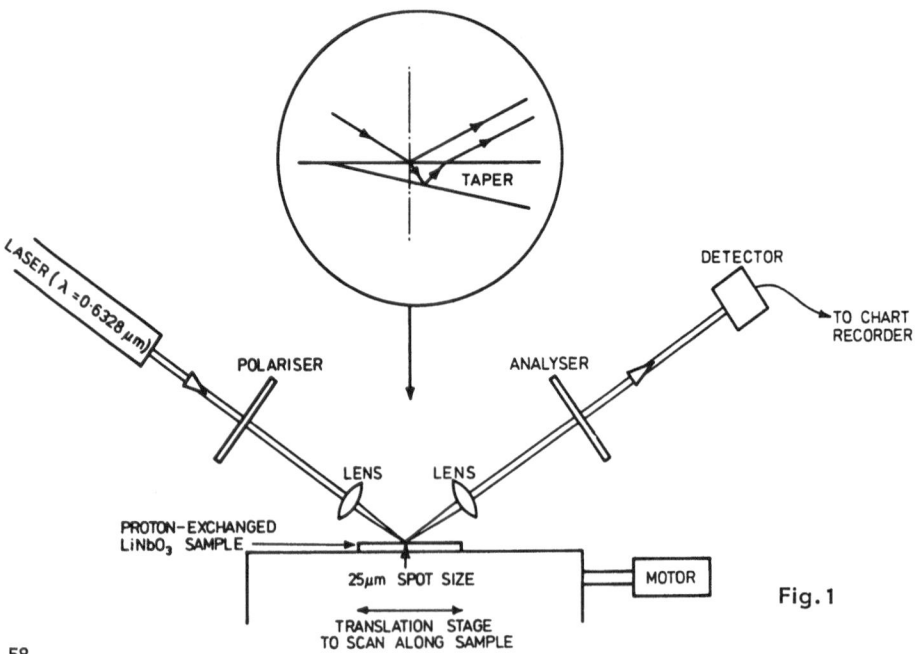

TAPER

LASER (λ =0.6328 μm)

DETECTOR

TO CHART RECORDER

POLARISER

ANALYSER

LENS   LENS

PROTON-EXCHANGED
$LiNbO_3$ SAMPLE

25μm SPOT SIZE

MOTOR

Fig. 1

TRANSLATION STAGE
TO SCAN ALONG SAMPLE

In order to analyse the tapers, the set-up illustrated in Fig. 1 was used. The output from the laser (λ = 0.6328 μm) is first passed through a polariser and then focussed to a 25 μm spot size. The reflected beam, consisting of reflections from upper and lower surfaces of the proton-exchanged layer,is passed through a second polariser and then to the detector. Interference between the two reflected components results in variation in the beam intensity when the beam is scanned along the taper region.

For an incidence angle of 70° and with values for film (n_f) and substrate index (n_s) of 2.3295 and 2.2025, it can be shown that the reflectivity is given by the expression:

$$R = \frac{0.525 - 0.049 \cos 2\phi}{1.0 - 0.049 \cos 2\phi} \qquad \ldots\ldots\ldots(1)$$

with $\phi = \dfrac{2\pi}{\lambda} n_f d \cos\theta_f = 3.368 \ (2\pi d)$ $\qquad \ldots\ldots\ldots(2)$

Fig. 2 shows experimental results obtained from four samples which were exchanged at T = 202°C for times of (a) 15 mins (b) 1 hour (c) 2 hrs 15 mins (d) 3 hrs 20 mins. The interference fringes indicate the form and extent of the tapered region. Fringes are only obtained with TE polarisation of the input beam as shown in (b) and (d) of Fig. 2. Table 1 below shows values of thicknesses computed from equation (2) compared with thicknesses obtained from waveguiding measurements,

P.E. time	w/g measurements	reflectivity measurement
15 min	0.28 μm	$\leq$ 0.3 μm
30 min	0.36	0.37
1 hr	0.677	0.6
2 hr 15 min	0.97	0.97
3 hr 20 min	1.14	1.12
6 hr 15 min	1.65	1.65

3. Application for coupling of waveguides

Previous work[2] has explored the possibility of constructing hybrid integrated optical systems using directional coupling techniques to interconnect waveguides in different materials. Coupling between glass and LiNbO$_3$ guides (with As$_2$S$_3$ overlays) has been demonstrated, but difficulties were encountered due to the need for exact phase-matching of the guided modes. Proton exchange, however, will allow taper-velocity coupling methods to be employed. Furthermore, proton exchanged waveguides and tapers may be formed over existing Ti-diffused guides (TIPE guides),thus allowing coupling to existing devices formed in the Ti-diffused guide.

For coupling between an As$_2$S$_3$ overlay guide on glass and a proton exchanged guide in LiNbO$_3$ the coupling coefficient, K, may be shown to be:

$$K \approx 0.23 \exp(-10.7\ c) \qquad \ldots\ldots\ldots(3)$$

where c is the interguide spacing and λ = 1.15 μm.

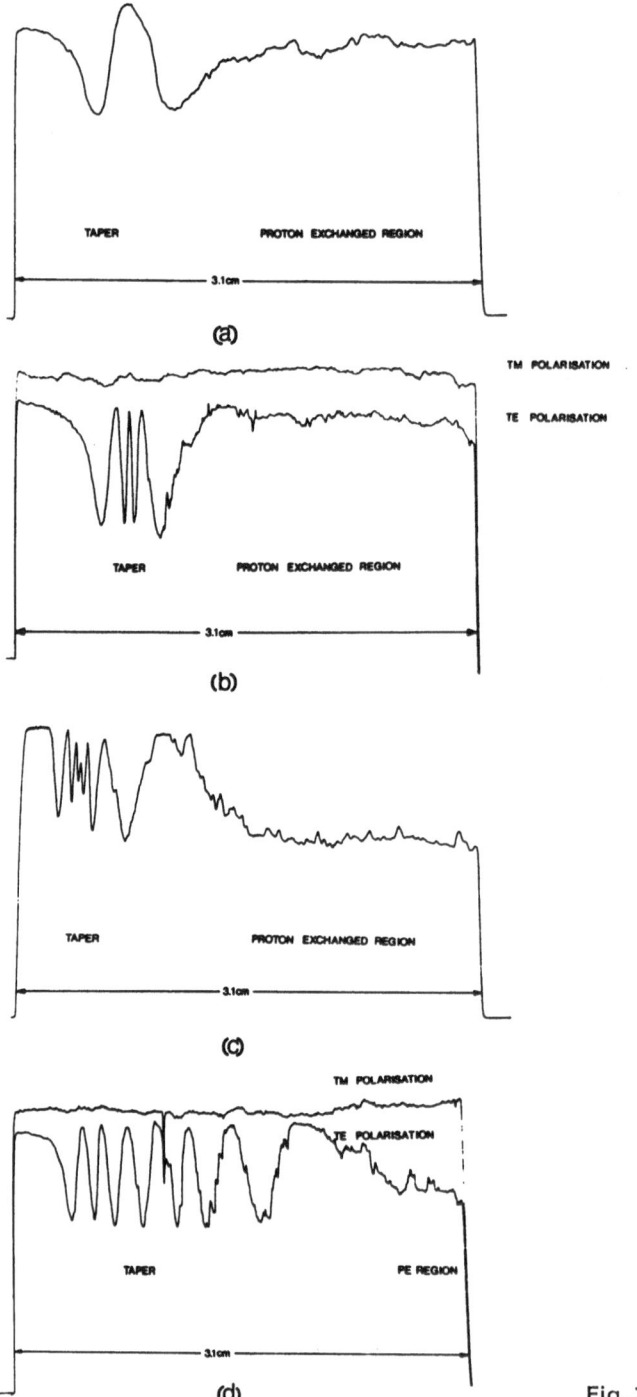

(a)

TM POLARISATION

TE POLARISATION

TAPER

PROTON EXCHANGED REGION

3.1cm

(b)

(c)

TM POLARISATION

TE POLARISATION

TAPER

PE REGION

3.1cm

(d)

Fig. 2

For efficient coupling, the taper length, ℓ , must satisfy
the condition[3]:

$$\ell > 213/\Delta\beta \qquad\qquad\qquad \dots\dots\dots(4)$$

where $\Delta\beta$ is the variation in propagation constant over the
taper length.

With $\Delta\beta \simeq \dfrac{2\pi}{\lambda}$ (0.1) we thus have $\ell > 0.4$ mm ($\lambda = 1.15\,\mu m$).

Hence, convenient taper lengths of 1/2 - 1 mm. are suitable for
coupling.

4. Conclusion
We have demonstrated a simple method for the formation of
waveguide tapers in $LiNbO_3$ substrates using proton exchange.
The tapers will allow directional coupling techniques to be
employed in the construction of hybrid integrated optical
systems.   Furthermore, a simple optical monitor set-up has been
designed for analysis of the tapers.   The optical monitor may
also be used for independent measurement of PE waveguide
thickness,   assessing   uniformity   of   PE   layers   and
characterisation of variable thickness layers.

Acknowledgements
The authors wish to acknowledge the Royal Society of
Edinburgh (G.S.) and the SERC (A.C.G.N.) for financial support

References

1. Nutt, A.C.G., et al. 2nd Europ. Conf. on Int. Opt., Florence,
   Oct. 1983.
2. Stewart, G., et al., 3rd Int. Conf. on Int. Opt. and Opt.
   Fibre Commun., San Francisco, April, 1981.
3. Milton, A.F., et al., Appl. Opt., Vol. 14, p. 1207, 1975.

Optical Damage Effects in LiNbO$_3$:Ti Waveguides

J.P. Nisius, P. Hertel, E. Krätzig, and H. Pape
Fachbereich Physik, Universität Osnabrück, Postfach 4469
D-4500 Osnabrück, Fed. Rep. of Germany

Holographic measurements of light induced refractive index changes are reported indicating stabilization of Fe^{2+} centers in LiNbO$_3$:Ti waveguides. These centers increase the sensitivity to optical damage effects.

1. Introduction

Light-induced refractive index changes [1]-socalled 'optical damage' effects - often impair the quality of optical components on the base of LiNbO$_3$. The effects are caused by impurities, which occur in different valence states in LiNbO$_3$ and which are present even in undoped (nominally pure) crystals. Especially effective are iron ions existing as Fe^{2+} and Fe^{3+} [2]. The concentration ratio $c_{Fe^{2+}}/c_{Fe^{3+}}$ can be varied in a wide range by heat treatments in special atmospheres. This ratio greatly influences the light-induced charge transport properties [3].

LiNbO$_3$:Ti waveguides are more sensitive with respect to optical damage than the bulk material, though Ti ions are found as Ti^{4+} only [4]. For this reason it is obvious to assume that the indiffusion of Ti^{4+} ions affects the valence states of the impurities, e.g. Fe ions.

In this contribution we investigate the Fe^{2+} concentration $c_{Fe^{2+}}$ as a function of the distance from the waveguide surface. This is performed by holographic means, utilizing the different intensity distributions of various modes.

2. Experimental Methods

The waveguides are fabricated by Ti indiffusion into y-cut LiNbO$_3$ substrates. The crystals with the Ti layer of e.g. 75nm thickness are heated to a temperature of 1000°C for 7h in argon atmosphere and cooled down in oxygen atmosphere for reoxidation. The waveguide chosen for this investigation can carry four TE and three TM modes. For the measurements we utilize the TM modes.

Analyses indicate that Fe ions are the dominant impurities in our LiNbO$_3$ substrate. The entire Fe concentration is determined in the oxidized substrate by EPR measurements to be 10 ppm ± 20%.

The photorefractive index changes are determined by measuring the diffraction efficiency of holographic gratings. Because the index changes are difficult to evaluate during the recording process, we investigate the optical

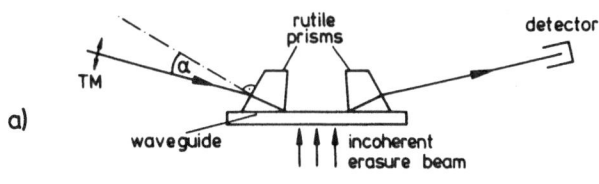

a)

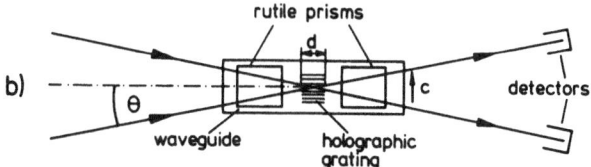

b)

Fig. 1

Experimental setup
(schematically)

a) front view

b) top view

erasure of the grating during homogeneous illumination with an incoherent beam, as shown in Fig. 1. We use light of an argon ion laser at a wavelength λ = 514 nm. The interaction length d is 7 mm and the angle 2θ between the two intersecting beams is 58 mrad.

3. Theoretical considerations

We denote by y the depth under the waveguide surface. Modes propagating along the x-axis are either TE or TM.

The propagation constants β_m and the mode profiles $E_m(y)$ resp. $H_m(y)$ are calculated as follows. Ti layer thickness and diffusion time allow to calculate the titanium concentration profile. Using the relation between concentration and refractive index change of ref. [5] we obtain the index profiles $\Delta n^e(y)$ and $\Delta n^o(y)$ which are Gauss curves of different width. The mode equations are solved numerically.

We work with modes that propagate under angles $\pm\theta$ with respect to the x-axis. Since θ is very small we may write

$$E_m^\pm(x,y,z) = E_m(y) e^{i\beta_m x\cos\theta} e^{\pm i\beta_m z\sin\theta} . \qquad (1)$$

For larger angles hybrid modes will enter the game.

By superimposing two such off-axis modes we generate a sinusoidal light intensity pattern:

$$|E_m^+(x,y,z) + E_m^-(x,y,z)|^2 \sim |E_m(y)|^2 \{1 + \cos K_m z\} \qquad (2)$$

$K_m = 2\beta_m \sin\theta$ is the grating vector. This light pattern, via the photorefractive effect, induces a change in both refractive indices which we shall denote by $\delta n^{o,e}(y,z)$.

In our case the saturation index change is dominated by the photovoltaic effect [6]:

$$\delta n(y,z) = \delta n_{sat}(y) \cos K_m z \text{ with } \delta n_{sat} = \frac{1}{2} n^3 r \frac{\kappa_0}{\kappa_1} \qquad (3)$$

r and n are the appropriate electrooptic coefficients and refractive indeces, respectively. κ_0 denotes the photovoltaic constant, κ_1 the specific photoconductivity. κ_0 depends only weakly on Fe concentration, and $1/\kappa_1$ is proportional to the concentration of Fe^{3+} centers [3]. Since our waveguides are highly oxidized we expect $\delta n_{sat}^{o,e}(y)$ to be nearly constant.

After saturation has been reached, the hologram is erased by uniform illumination with incoherent light. The diffraction efficiencies are monitored. Calculation shows that for our experiment coupling from $E_m^\pm$ to $E_n^\pm$ is negligible (phase mismatch). Thus, the diffraction efficiency η_m for coupling from $E_m^\pm$ to E_m^- or vice versa is of interest only.

The mode equation without light-induced refractive index change can be written in the form

$$W_m E_m^a = 0 \quad \text{with } a = +, -. \tag{4}$$

With light-induced index change one has to solve

$$\{W_m + k_0^2 \, 2n \, \delta n(y) \cos K_m z\} \, E = 0 . \tag{5}$$

We adopt Kogelnik's method [7] and write

$$E(x,y,z) = \sum_b c_m^{ba}(x) \, E_m^b(x,y,z) \quad \text{where} \tag{6}$$

$$c_m^{ba}(0) = \delta_{ba} \tag{7}$$

describes the initial condition: at $x = 0$ the field is $E_m^a(0,y,z)$. Higher diffraction orders and second derivatives of $c_m^{ba}(x)$ are neglected. Multiplying with $E_m(y)$ and integrating yields

$$c_m^{++}(x) = c_m^{--}(x) = \cos \frac{\pi x}{\lambda} <\delta n>_m ,$$
$$c_m^{+-}(x) = c_m^{-+}(x) = i \sin \frac{\pi x}{\lambda} <\delta n>_m . \tag{8}$$

Here the average value

$$<\delta n>_m^e = \frac{\int dy \, |E_m(y)|^2 \, \delta n^e(y)}{\int dy \, |E_m(y)|^2} \tag{9}$$

refers to TE modes, TM modes are handled analogously (ordinary refractive index change, $H_m(y)$ instead of $E_m(y)$). We thus obtain Kogelnik's formula

$$\eta_m = |c_m^{-+}(d)|^2 = \sin^2(\frac{\pi d}{\lambda} <\delta n>_m) \tag{10}$$

for small θ with an averaged light-induced refractive index change. Note that $<\delta n_{sat}>_m$ should be approximately the same for different m. The TM analogue to (9) requires $\Delta n^o >> \delta n^o$.

The time-dependence of $\delta n(y)$ during erasure is approximately exponential [8,9]:

$$\delta n(y) = \delta n_{sat} \, \exp[-\sigma^{ph}(y) t/\varepsilon \varepsilon_0] . \tag{11}$$

We may safely neglect the dark conductivity. It follows that for short enough times we have

$$<\delta n>_m = \delta n_{sat}[1-t<\sigma^{ph}>_m/\varepsilon\varepsilon_o+...] \ . \tag{12}$$

The photoconductivity in this case is usually expressed with the help of the absorption constant α and the light intensity I: $\sigma^{ph} = \kappa_1\alpha I$. We erase with homogeneous light, and κ_1 will only weakly depend on y. Since absorption in the 500 nm region is mainly caused by Fe^{2+} centers [2], one may calculate $<c_{Fe^{2+}}>_m$ from the measured $<\sigma^{ph}>_m$ values. We contrast these averages with the penetration depth of the corresponding mode

$$<y>_m = \frac{\int dy |E_m(y)|^2 y}{\int dy |E_m(y)|^2} \ . \tag{13}$$

4. Experimental Results and Discussion

The assumption of a nearly constant saturation value of δn for the various modes - which is equivalent to a nearly spatially constant Fe^{3+} concentration - is confirmed by the experiments. We obtain with (10) from the measured saturation values of n_m:

$$<\delta n^o_{sat}>_{TM0} = 2.4\cdot10^{-5}, <\delta n^o_{sat}>_{TM1} = 1.9\cdot10^{-5}, <\delta n^o_{sat}>_{TM2} = 1.6\cdot10^{-5}.$$

A quantity δn_{sat} constant in y yields, according to (3), a quantity κ_1 constant in y. With a photovoltaic constant [3] $\kappa_o = 5\cdot10^{-12}$m/V we deduce from our data κ_1 to be approximately $1.5\cdot10^{-17}m^2/V^2$.

From the erasure curves we obtain with (12):

$$<\sigma^{ph}>_{TM0}=4.8\cdot10^{-17}A/(Vm), <\sigma^{ph}>_{TM1}=0.91\cdot10^{-17}A/(Vm), <\sigma^{ph}>_{TM2}=0.58\cdot10^{-17}A/(Vm).$$

The photoconductivities refer to a light intensity of 1 W/m^2.

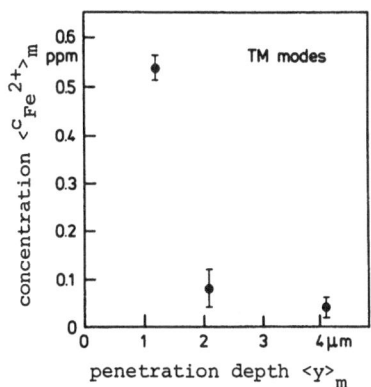

Fig. 2
Fe^{2+} concentration $<c_{Fe^{2+}}>_m$ determined from optical erasure measurements versus penetration depth $<y>_m$

Utilizing the results of ref. 2 we finally obtain the Fe^{2+} concentrations which are presented in Fig. 2 as a function of the penetration depth $<y>_m$ for the three TM-modes. - Similar results-strong enhancements of the $c_{Fe^{2+}}$ values near the surface- have been found for other waveguides.

The indiffusion of Ti ions may introduce additional Fe ions into the waveguide because the Ti used for vacuum deposition contains impurities, above all Fe. Furthermore, there might be a migration of Fe ions to the surface during the diffusion process. From the measured $<\delta n_{sat}>_m$ values we conclude that an increase in the entire Fe concentration up to 50% might be possible near the surface.

Anyhow, in oxidized crystals the concentration ratio $c_{Fe^{2+}}/c_{Fe^{3+}}$ increases towards the surface: Fe^{2+} centers are stabilized in the region of large Ti concentration.

References

1 A. Ashkin, G.D. Boyd, J.M. Dziedzic, R.G. Smith, A.A. Ballman, J.J. Levinstein, and K. Nassau: Appl. Phys. Lett. 9, 72(1966)

2 H. Kurz, E. Krätzig, W. Keune, H. Engelmann, U. Gonser, B. Dischler, and A. Räuber: Appl. Phys. 12, 355(1977)

3 E. Krätzig: Ferroelectrics 21, 635(1978)

4 A.M. Glass, J.P. Kaminov, A.A. Ballman, and D.H. Olson: Appl. Opt. 19, 276(1980)

5 J. Vollmer, J.P. Nisius, P. Hertel, and E. Krätzig: Appl. Phys. A32, 125(1983)

6 A.M. Glass: Opt. Eng. 17, 470(1978)

7 H. Kogelnik: Bell Syst. Techn. J. 48, 2909(1969)

8 J.J. Amodei: RCA Rev. 32, 185(1971)

9 H. Kurz: Optica Acta 24, 463(1977)

Direct Measurement of Refractive Index Profiles of Ti:LiNbO₃ Slab Waveguides

A. Neyer

Universität Dortmund, Lehrstuhl für Hochfrequenztechnik, Postfach 500500
D-4600 Dortmund 50, Fed. Rep. of Germany

The reflectivity measurement of angular polished surfaces is utilized to
determine directly the refractive index profiles of Ti:LiNbO₃ slab waveguides with a resolution of $\Delta n/n = 10^{-4}$.

1. Introduction

It is of fundamental importance for the design and reliable fabrication of
integrated optical circuits to have an accurate knowledge about the refractive index profiles of the fabricated waveguides. There exist already a
number of methods to determine these profiles indirectly from concentration
measurements of the dopants /1,2/. However, nonlinear dependences of the
concentrations with the index changes, as in the case of titanium diffusion
into LiNbO₃ /2/, as well as the polarization dependence and the wavelength
dependence of the index profiles on the concentration makes the precise
evaluation of the index profiles more difficult.

In this paper we present a direct method to determine the refractive
index profiles of Ti:LiNbO₃ slab waveguides. This method, which is based on
the reflectivity measurement of angular polished waveguide surfaces, has
been adopted from the index profile measurements of ion-exchanged glass
waveguides /3/ and is applied here, for the first time, to Ti:LiNbO₃ waveguides.

2. Principle of Reflectivity Profiling

The experimental set-up for measuring the index profiles of integrated
optical slab waveguides is shown in Fig. 1. The angular polished surface of
the sample is illuminated by monochromatic, polarized light at normal incidence by means of an incident light microscope. The reflected light power
is detected by the array vidicon on top of the microscope. The vidicon is
part of an optical multichannel analyser system (OMA 2, PAR). The system is
programmed to subdivide the magnified image (20x) of the sample into 500
parallel stripes, corresponding to the number of available channels. By

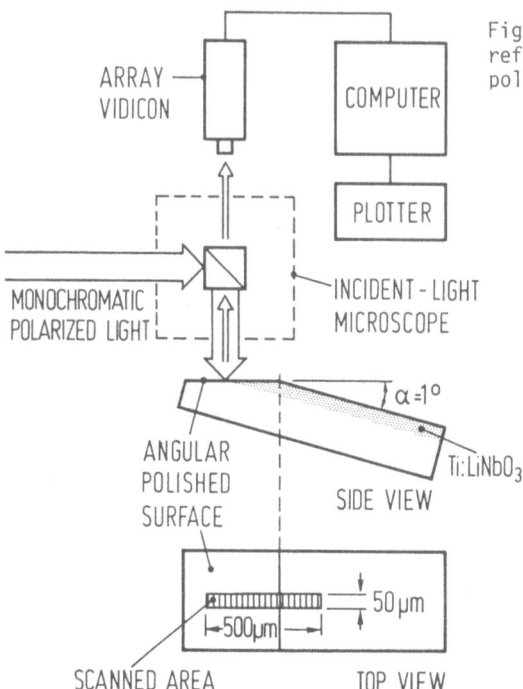

Fig. 1. Experimental set-up for
reflectivity profiling on angular
polished surfaces

means of apertures, the imaged area of the sample is limited to 500 μm × 50 μm
so that a stripe of 1 μm × 50 μm is attributed to each channel. Thus, the
resulting resolution in depth is $1.7 \cdot 10^{-2}$ μm at an polished angle of 1°.
The change in the reflectivity $\Delta R = R - R_0$, where R is the reflectivity of
the indiffused material and where R_0 is that of the substrate, is directly
related to the change in the refractive index Δn by

$$\frac{\Delta n}{n_0} \simeq \frac{1}{4} \left(n_0 - \frac{1}{n_0} \right) \frac{\Delta R}{R_0}$$

where n_0 is the refractive index of the substrate material. This simplified
formula is an excellent approximation up to reflectivity changes $\Delta R/R_0$ of
10 %.

3. Experimental Results

The basic problem of the here described method of index profile measurements
is the precise angular polishing procedure, especially at small angles of
about 1°. The requirements to the polished surfaces are: flatness better
than λ/4, roughness smaller than ± 20 nm and no rounding effects at the
edge which would lead to errors in the determined index change at the sur-

POLISHED SURFACE

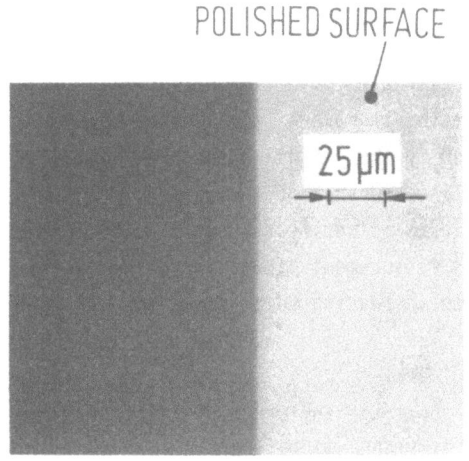

Fig. 2. Nomarsky interference contrast photograph of an angular polished surface

face. Good results are obtained with lead-tin polishing plates which are specially turned off, and diamond spray as polishing medium. Fig. 2 shows the polished surface of Ti:LiNbO$_3$ as seen by a Nomarsky contrast microscope. The rounded region at the edge is about 5 µm wide, resulting in an error for the first 0.1 µm of the index profile. The homogeneous grey-shade of the polished surface indicates that the flatness requirements are fulfilled. The residual surface roughness, not visible in the microscope, is so low, that relative reflectivity changes of $0.4 \cdot 10^{-3}$ could be detected, permitting a resolution of the relative index change of about 10^{-4}. Assuming a maximum relative index change of 10^{-2} in Ti:LiNbO$_3$-waveguides, the resolution of an index profile by the proposed method is in the order of 1 %.

Fig. 3 shows the measured refractive index profile of a Ti:LiNbO$_3$ slab waveguide at λ = 633 nm. The LiNbO$_3$ substrate is Y-cut and the measured

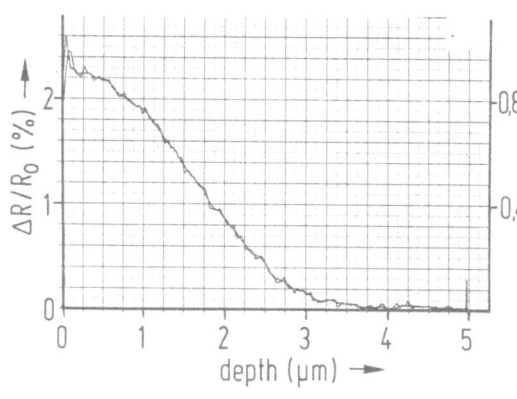

Fig. 3. Measured reflectivity and refractive index profile of a Ti:LiNbO$_3$ slab waveguide (data in the text) at λ = 633 nm

curve shows the relative index change of the extraordinary index $\Delta n_e/n_e$. The fabrication parameters of the waveguide are: Ti-thickness: 500 nm, diffusion temperature: 1000°C, diffusion time: 9 h. The diffusion was carried out in water vapor-enriched oxygen atmosphere, whereby the water bubbler was heated to 40°C. As predicted by the diffusion theory and the rather linear relationship between Ti-concentration and the induced change of n_e /3/, the measured index profile coincides nearly exactly with a Gaussian function. However, previous measurements indicate that the shape of the index profile will change with increasing temperature of the water bubbler.

More detailed results, especially about the influence of the water vapor on the refractive index profiles of Ti:LiNbO$_3$ waveguides, as well as the wavelength and polarization dependence of the profiles, will be given at the conference.

4. Conclusion

The reflectivity measurement of angular polished surfaces is utilized to determine directly the refractive index profiles of Ti:LiNbO$_3$ waveguides. For this purpose, a special polishing technique has been developed to produce very flat surfaces with a very low roughness. First measurements demonstrate that the index profiles can be resolved with an accuracy of 1 % at a maximum relative index change of about 1 %.

5. Acknowledgement

The author would like to acknowledge the expert polishing of the samples by Mr. D. Dettmar and the kind support at the profile measurements by Dr. R.F. Heidemann.

6. References

/1/ K. Sugii, M. Fukuma and H. Iwasaki, J. Mater. Sci. 13 (1978),
   pp. 523-533
/2/ M. Minakata, S. Saito and M. Shibata, J. Appl. Phys. 50 (1979),
   pp.3063-3067
/3/ H.-J. Lilienhof, K.F. Heidemann, D. Ritter and E. Voges, Optics Comm.
   35 (1980), pp. 49-53

Buried Single-Mode Channel Waveguides in BK 7 by Field Assisted Cs-Ion Exchange

H.-J. Lilienhof and H.W. Hölscher

Universität Dortmund, Lehrstuhl für Hochfrequenztechnik, Postfach 500500
D-4600 Dortmund 50, Fed. Rep. of Germany

Buried single mode channel waveguides are fabricated in BK 7 by a masked, field assisted Cs-ion exchange followed by an unmasked thermal Na-ion exchange.

1. Introduction

For telecommunication with monomode fiber-optic systems optical branching networks are requested. Optical data distribution requires low-loss single-mode stripe waveguides with optimized index profiles for an efficient fiber coupling /1/. Furthermore, single-mode channel waveguides can be applied as optical sensors, e.g. for structures like ring resonators in integrated optical rotation sensors /2/. For the reason of cheap mass fabrication these devices should be constructed in low-cost materials, such as glasses.

Thermal ion exchange is the most common process to fabricate single-mode waveguides in glass. This technique, in combination with the photolithographic technique, offers the advantages of reproducibility and batch fabrication. The most popular exchange process until now is the silver-ion exchange in ordinary borosilicate or sodalime glasses, to a lesser extent high quality optical glasses such as BK 7 have been used /3/. However, the silver-ion exchange exhibits problems which can be attributed to the tendency of silver-ions to reduce to colloidal metal, and to segregate to metallic particles. This leads to strong scattering losses, in particular if metal masks are used for the delineation of channel waveguides. In addition, the presence of silver ions in glass leads to an increased intrinsic absorption.

It is well known that ion species such as cesium or thallium do not increase the intrinsic losses of glasses. Tl-ions have been used to fabricate multimode branching structures in glass /4,5/. However, the use of Tl-melts arises problems due to the toxicity of thallium.

2. Field-Assisted Fabrication of Waveguides

We used Cs-ions to fabricate single mode stripe waveguides in an optical glass (BK 7, Fa. Schott, Mainz), and utilized a field-assisted exchange

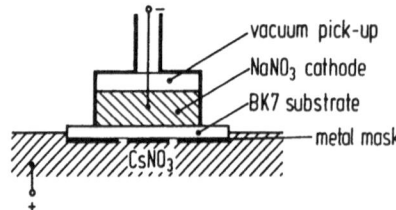

Fig. 1. Configuration of a masked and
field-assisted cesium-ion exchange for
optical stripe waveguides

process. This has the advantages of profile shaping for an efficient
coupling to monomode fibers by suitable technological parameters and of
short (in the order of minutes) exchange times.

The geometry of a masked Cs-ion exchange for optical stripe waveguides is
depicted in Fig. 1. Slides of BK 7 glasses (1 mm thick) are prepared with
the aid of conventional photolithographic techniques. Stripe waveguides
are defined by etching windows into a masking material (2000 Å Al or Ti)
which is used as a barrier to the ion exchange process. Thorough cleaning
and prebaking of the substrates are necessary to achieve metal layers
which are free from pin-holes. The cathode consists of pure $NaNO_3$-melt. The
substrate is attached to a vacuum pick-up, and is inserted into an oven
with a precision temperature control. After a slow preheating to the desired temperature, the masked surface is dipped into the $CsNO_3$-melt. For
single-mode stripe waveguides at 820 nm we used the following parameters:

temperature: 720 K
time: 2 min
applied voltage: 50 V
width x_w of mask windows: 3 μm .

For low losses it is important to reduce the surface scattering by burying
the waveguides below the surface. For this purpose we used a second unmasked exchange in a pure $NaNO_3$-melt. This process is a thermal one. Typical parameters are

temperature: 720 K
time: 30 min .

For most applications the waveguides can adequately be characterized by two
parameters, near-field distribution and attenuation.

3. Measurements of the Near Field Pattern

The experimental arrangement for the measurements of the near-field distribution of the stripe waveguides is shown in Fig. 2. By means of a monomode optical fiber and a cladding modes stripper, the light of a Hitachi

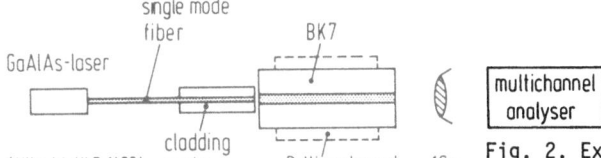

Fig. 2. Experimental set-up for the measurements of the near field distribution

HLP 1400 GaAlAs-laser (wavelength 820 nm) is butt-coupled into the wave-guides. With the help of a microscope objective (magnification: 16x) the output light pattern is projected on the diode array of an optical multi-channel analyser. The spatial resolution of the apparatus is limited by the resolution of the objective to about 0.3 µm.

4. Experimental Results

The measured near-field distributions of a Cs-ion exchanged stripe wave-guide are shown in Fig. 3. In the left panel (Fig. 3a) the almost Gaussian field profile parallel to the surface is shown. Contrary to this symmetric profile the in-depth field profile in the right panel is quite asymmetric due to the surface (Fig. 3b). The slow decrease of the intensity on the substrate side of the waveguide results from the scattering of the guided light at the surface associated with undesirable high losses.

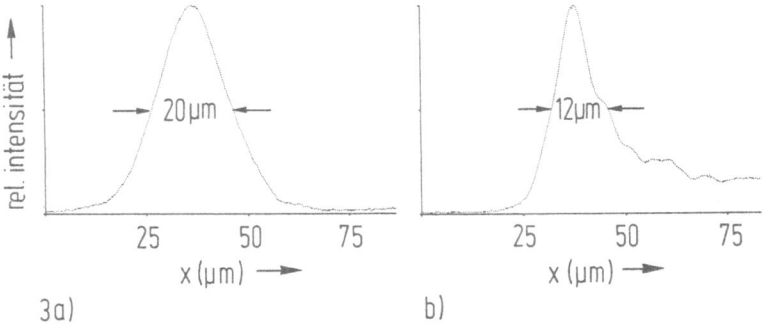

Fig. 3 a,b. Near field pattern of a Cs-ion exchanged waveguide at λ = 820 nm, (a) parallel to the surface, (b) in depth, exchange parameters: T = 720 K, t = 2 min, U = 50 V, x_w = 3 µm

To reduce the influence of the surface, the waveguides were buried in the way described above. The change of the near-field distribution of the same waveguide, shown in Fig. 3, is demonstrated in Fig. 4. The horizontal and in-depth field profiles are strongly changed. The field pattern is almost circular, and it has a reduced diameter.

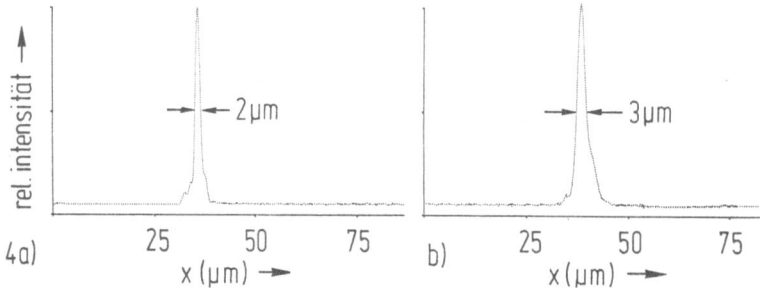

Fig. 4 a,b. Near field pattern of a buried waveguide at λ = 820 nm, (a) parallel to the surface, (b) in depth, exchange parameters : as in Fig. 3 and T = 720 K, t = 30 min

Due to small scattering losses of the waveguides,the attenuation cannot be measured accuarately by usual techniques (e.g. sliding-prism-coupler, three-prism-coupler). For low-loss waveguides - as fabricated here - only resonator measurements are useful /6/. Results for ring resonators will be presented at the conference.

5. Acknowledgement

This work has been performed for Bodenseewerk Gerätetechnik GmbH, and has been supported by the BMFT.

6. References

/1/ W.K. Burns, G.B. Hocker, "Endfire coupling between optical fibers and diffused channel waveguides", J. Appl. Opt. 16,8 (1977), 2048

/2/ K. Honda, E. Garmire, K. Wilson, "Characteristics of an integrated optics ring resonator fabricated in glass", J. of Lightwave Technology LT-2, 5 (1984), 714

/3/ R.G. Eguchi, E.A. Maunders, I.K. Naik, "Fabrication of low-loss waveguides in BK 7 by ion exchange", Proceedings SPIE 408 (1983)! 21

/4/ M. Hafich, D. Chen, J. Huber, "Properties of optical waveguides formed by thermal migration of thallium ions in glass", Appl. Phys. Lett. 33 (1978), 997

/5/ E. Okuda, I. Tanaka, T. Yamasaki, "Planar gradient-index glass waveguide and its applications to a 4-port branched circuit and star coupler", 4th Topical Meeting on Gradient Index Optical Imaging Systems (1983)

/6/ R. Regener, W. Sohler, "Loss in low finesse Ti:LiNbO₃ optical waveguide resonators", to be published in Appl. Phys. A (1985)

Fabrication and Characterization of Buried Glass Waveguides with Symmetric Index Profile

R.K. Lagu, V. Ramaswamy, and S.I. Najafi

Department of Electrical Engineering, University of Florida
Gainesville, FL 32611, USA

Diffusion studies and modal behavior of Ag^+-Na^+ exchanged planar waveguides in glass substrate are presented. In addition, a simple SEM technique to determine the refractive index profile and the realization of symmetrical buried waveguides with fiber-like refractive index profile are also reported.

1. Introduction

Ion-exchanged glass waveguides have received considerable attention recently and they are expected to form the basis for passive integrated optical components. Significant progress has been made in the fabrication of ion-exchanged waveguides. Recent electrolytic release technique [1] and the improved method [2] using fritted glass bulb around the cathode provide opportunities for fabrication of passive waveguide devices with repeatable characteristics. This is quite important because glass waveguide devices being passive, cannot be tuned electro-optically to compensate for fabrication errors.

The guiding characteristic of a planar waveguide depends on the device parameters such as maximum index change, index profile, effective guide depth, which in turn depend on the process parameters like the diffusion temperature, time and silver ion concentration in the bath. To design a waveguide supporting specific number of modes with desired propagation constants, it is necessary to determine relations between process and device parameters. In this paper, we present an empirical relation between process and device parameters. We also present a procedure for fabricating symmetrical buried waveguides which are comparable in shape and size to the core of single mode fibers. A precise technique to characterize these waveguides using backscattered electrons in SEM is also reported.

2. Planar Surface Waveguide Diffusion Characteristics

The Ag^+-Na^+ ion-exchanged waveguides are fabricated in a molten $NaNO_3$ bath with small concentration of silver ions. The bath can be considered to be an infinite source of silver ions as the number of ions diffusing into the glass sample is very small compared to that of the ions present in the bath. Under these conditions, the diffusion equation has a closed form analytical solution, namely the complementary error function. Planar waveguides fabricated in this manner, therefore, exhibit a complementary error function index profile. The b-V characteristics for this highly asymmetric profile have been determined by solving the normalized mode dispersion equation [3] and the cut-off values for the V parameter for the first three modes are found to be 2.7, 6.3 and 9.9.

The V parameter is defined as

$$V \approx \frac{2\pi}{\lambda} d \sqrt{2n_b \, \Delta n}$$

where λ is the wavelength of operation, d is the effective waveguide depth measured between values at 0.168 of peak refractive index at the surface, and is given by $d = 2\sqrt{Dt}$ where t is the time of diffusion and D is the diffusion coefficient, n_b is bulk refractive index, and Δn is the maximum refractive index change which depends on the concentration of silver ions C_0.

We fabricated several sets of single and multimode waveguides using Labmate microscope slides for different value of C_0 and measured the propagation constants of the modes they suported. Using an iterative computer program, the data was fitted into the b-V characteristics of the erfc profile to estimate Δn and d. Using these results, the value of D was estimated and the relation between Δn and C_0 plotted as shown in figure 1. It was noted the diffusion coefficient D did not vary much with different brands (e.g., Corning, Schott and Labmate) of glass substrates.

For small C_0, Δn versus C_0 is essentially linear. For guides with small number of modes, the relationship between Δn and C_0 can therefore be approxi-

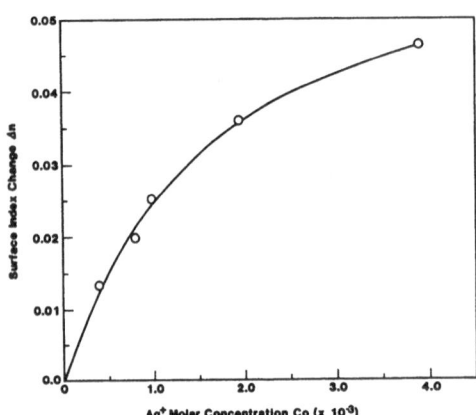

Figure 1. Relation between sur-
face index change Δn and Ag^+ concentration

mated as

$$\Delta n \approx 26.5 \; C_0$$

C_0 is the mole fraction of silver ions in bath.

Using above empirical relations between the device (d, n_b, Δn) and the process parameters (C_0, D, t), we can approximate the V parameter as

$$V = 63.85 \; \sqrt{C_0} \, t$$

for λ = 0.6328 μm, n_b = 1.512 and D = 0.129 μm^2/min.

Using cut-off values for V parameters for various modes, the design curves are plotted in t-C_0 plane as shown in figure 2.

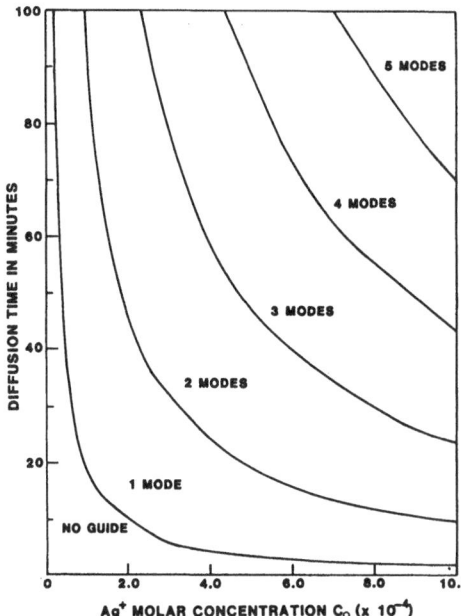

Figure 2. Design curves for various modes (λ = 0.6328 μm)

3. Waveguide Characterization

The technique frequently used for measuring silver ion concentration seems to be Electron Microprobe Analysis, also referred to as Electron Microprobe [4]. The equipment needed for this technique is quite expensive and is not available at our university. We have investigated a technique using a Scanning Electron Microscope to find the silver ion concentration profile. In an SEM, the specimen is bombarded by an accelerated electron beam with an accelerating voltage in the range of 20-50 Kv. The specimen generates two types of electrons, namely, the secondary electrons and the back-scattered electrons. The back-scattered electrons which have higher energies carry the

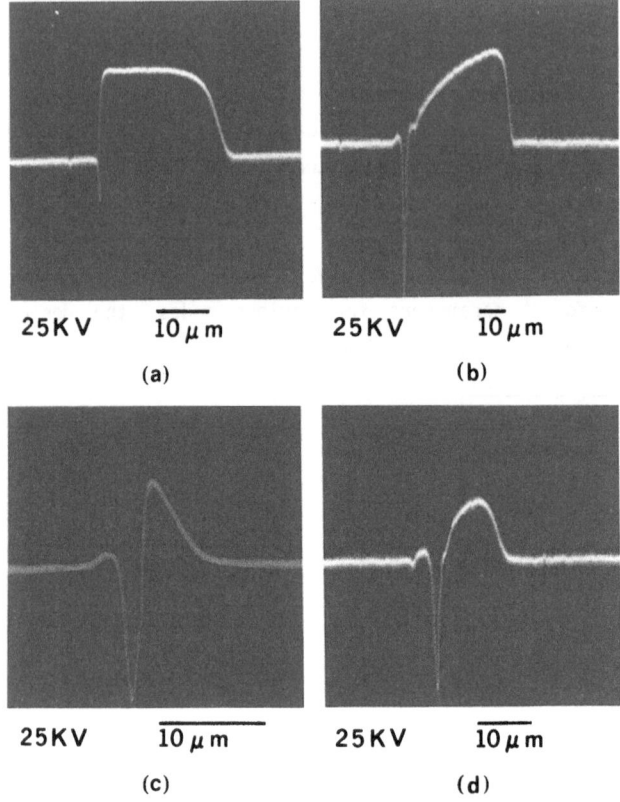

Figure 3. a) Index profile of a planar waveguide fabricated in the first step by field-assisted diffusion of silver ions. b) Second field-assisted diffusion in pure NaNO$_3$ results in a buried skewed index profile. c) Index profile of a planar waveguide fabricated in the first step by diffusion without applied electric field. d) Symmetrical index profile of a buried waveguide results of second field-assisted diffusion in pure NaNO$_3$.

information about specimen composition. Since the atomic weight of silver is much higher than that of the rest of the ions in the glass substrate, the contribution of silver ions in the guide region to the yield of back-scattered electrons is maximum. Thus, if the sample is polished across the edge of the waveguide and scanned in the SEM, the back-scattered electron intensity profile gives a good estimate of the silver ion concentration profile. Since the refractive index change is directly proportional to silver ion concentration, it also gives a good estimate of the index profile. The silver ion concentration profiles measured using this technique are shown in figures 3(a) through 3(d). It can be seen that this method has a good signal to noise ratio and high resolution. The big dip seen in the signal comes from a layer of epoxy used while polishing the waveguide. The epoxy is applied on the surface of the guide and another glass slide is stuck to it before polishing so as to avoid the rounding at the edge of the waveguide.

4. Planar Buried Waveguide Fabrication

A two step diffusion process utilizing field-assisted diffusion is used in glass to form buried waveguides [5]. In this process, soda-lime-silicate glass slides are immersed in molten sodium nitrate where silver ions are also present. Initially, a surface waveguide is formed by a field-assisted diffusion of silver ions in glass. This is followed by a second field assisted diffusion, however in pure sodium nitrate. The diffusion time and applied field in each of the two steps are two important parameters which control the dimensions and index profile of a buried waveguide.

In the existing process, applied field in the range of 60-100 volts/mm is used. This results in a step-like waveguide (figure 3(a)), which when buried, gives a skewed index profile (figure 3(b)). These results agree with the skewed profiles of buried waveguides already reported in literature [5].

We have experimentally determined that if no field is applied in the first step, as expected, a surface waveguide with complementary error function profile is formed (figure 3(c)) which, when subjected to a field-assisted diffusion in pure sodium nitrate, results in a symmetrical structure with near parabolic index profile as shown in figure 3(d). To obtain waveguides with various widths, surface waveguides were formed without drift field for 40 minutes using the electrolytic process [2]. The time of second diffusion was varied between 20 to 80 minutes using an electric field of 30 volts/mm. The resulting waveguides were polished and analyzed using a Scanning Electron Microscope (SEM) to get the estimate of their widths. Figure 4 shows a plot of waveguide width W (measured between 95% peak values) versus the second step diffusion time indicating that symmetrical buried waveguides of arbitrary

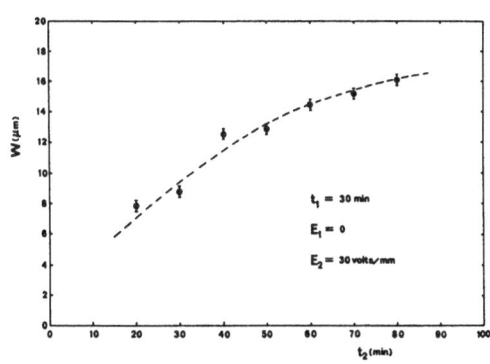

Figure 4. Relation between the waveguide width and the second step, field-assisted diffusion time

width in the range of 6-16 μms can easily be achieved and that diffusion depth exhibits the expected $\sqrt{t}$ dependence.

5. Conclusion

Design curves which can be used to determine the concentration of silver ions and the time of diffusion required for a planar waveguide of given width, number of modes, and surface index change are reported. A simple, new technique which uses the back-scattered electrons in an SEM to determine the refractive index profile and waveguide width of glass waveguides is also described. In addition, we have presented a procedure for fabricating buried glass waveguides with fiber-like index profile of varying widths.

Acknowledgments

The authors would like to thank E. J. Jenkins of the Department of Material Science and Engineering for his assistance in SEM analysis and C. Wayne Twiddy for his help in waveguide fabrication. This work was supported in part by a grant from AFOSR Contract number AFOSR 84-0369.

References

1. R. K. Lagu and R. V. Ramaswamy, "Fabrication of single mode glass waveguides by electrolytic release of silver ions," Applied Physics Letters, 45, 117-118, 1984.
2. S. I. Najafi, R. V. Ramaswamy and R. K. Lagu, "An Improved method for fabricating ion-exchanged waveguides through electrolytic release of silver ions," to be published in IEEE J. Lightwave Tech.
3. R. V. Ramaswamy and R. K. Lagu, "Numerical field solution for an arbitrary asymmetrical graded-index planar waveguide," IEEE J. Lightwave Tech. JLT-1, 408-417, 1983.
4. G. Stewart, C. A. Miller, P. J. R. Laybourn, C. D. W. Wilkinson and R. M. Delaru, "Planar optical waveguides formed by silver ion migration in glass," IEEE J. Quantum Electron., QE-13, 192-200, 1977.
5. G. Chartier, P. Collier, A. Guez, P. Jaussaud and Y. Won, "Graded index surface or buried waveguides by ion-exchange in glass," Applied Optics, 19, 1092-1095, 1980.

Part III

Semiconductor-Devices

Progress in Integrated Optics Lasers

Yasuharu Suematsu

Department of Physical Electronics, Tokyo Institute of Technology
O-okayama, Meguro-ku, Tokyo 152, Japan

Recent progress in the integrated optics lasers especially for the dynamic-single-mode (DSM) semiconductor lasers with distributed structures are reviewed. The DSM laser is developed as light sources with pure spectrum even under the dynamic operation for the wide-band optical fiber communication in the lowest loss wavelength region of 1.5μm, in which the techniques in the integrated optics are utilized. It consists of a mode selective resonator and a transverse-mode-controlled waveguide, such as the narrow-striped distributed-Bragg-reflector (DBR) or distributed feedback (DFB), so as to maintain a fixed axial-mode under the rapid direct-modulation.

In an application of DSM lasers for the light sources in the 1.55μm single mode fiber transmission system, the normalized transmission bandwidth of 25Gbits/s $(km)^{1/2}$ is theoretically estimated in comparison with an experimental top data of 23Gbits/s$(km)^{1/2}$.

I. Introduction

In the 1.55μm wavelength single mode fiber communications in the minimum loss wavelength region of 1.55μm [1], the effect of chromatic dispersion [2] prevent sometimes the high capacity and long distance information transmissions when conventional laser diodes are used for those light sources. Because the spectral broadening is appears in conventional laser diodes, when it is modulated rapidly, say short pulse operation less than a few nanosecond [3],[4].

A single mode laser with the character of single-wavelength operation even under rapidly modulated condition in the wavelength of 1.5μm was first realized by an integrated twin-guide lasers with a distributed-Bragg-reflector (DBR) observed at the pulse width of 1.5nsec [5],[6]. This comes after some trials at the 1.3 μm wavelength region [7],[8]. Distributed feedback (DFB) lasers and other kinds of distributed Bragg reflector lasers were also developed for the same purpose [9]-[12]. Presently these lasers are called as "dynamic-single-mode (DSM) lasers" [13].

A DSM laser is tightly mode-controlled laser to select a fixed wavelength imposing additional losses to neighbouring sub-modes by a wavelength selective resonator, such as, the distributed reflector for the longitudinal mode control, the narrow stripe waveguide for the transverse mode control, and by the TE mode filter for the polarization control.

Studies on the wavelength-controlled laser diode reported, which made basis for the DSM lasers, were started somewhat earlier with the DFB laser [14],[15] in 1973 [16] to 1974 [17]-[21], the coupled cavity laser in 1974 [22] to 1975 [23], the DBR laser [24] in 1975 [25] to 1976 [26], and the short cavity lasers in 1979 [27]. A single mode operation of DFB lasers at the observed width of somewhat wider 5nsec pulse was reported [28]. Those works were done at the wavelength of 0.85μm. In the theoretical point of view, possibilities of single mode operation was discussed on DFB lasers with additional mirrors or tapering the pitch in the periodic corrugation were pointed out [29],[30]. To avoid some theoretical difficulty of single-mode operation in DFB lasers, DBR lasers were extensively studied [31].

After the development of low loss single mode fiber in the wavelength of 1.55μm, it became interested to realize the dynamic single mode lasers in this wavelength [13]. First demonstration was done by DBR type lasers, as mentioned above. Due to simpler fabrication, DFB lasers with an end mirror were extensively developed for this purpose. The C^3 laser has been also tested for its easiness in experiments [32]. The additional loss imposed to the neighbouring submodes to maintain the dynamic single-mode operation[33] was given theoretically [34], which gave a reference to select a proper structure for this purpose. The effect of end mirror in DFB lasers was studied [35],[36]. Various DFB lasers were also reported [37]-[42]. From such consideration it was more clear that the axial mode selectivity is more dominated for the DBR lasers or DFB lasers with phase adjusting coupling regions [43], therefore further research activity is toward more complicated and better performed new distributed reflector devices [44]-[46]. The technology of monolithic integration of optical circuits and devices has contributed to integrated additional resonator.

It was found that the lasing wavelength of DSM laser shifts periodically in one period of direct modulation [4],[47]. This dynamic wavelength shift gives an equivalent spectral width or the dynamic spectral width. The maximum value of the dynamic wavelength shift finally determines the transmission bandwidth of a conventional single-mode fiber[47]. DSM lasers with distributed reflector demonstrated not only for the purity of the lasing spectrum but also the higher power output [40], [48], [49] and the low noise characteristics of InGaAsP/InP DFB lasers[50].

In this paper, the progress of dynamic-single-mode (DSM) lasers is reviewed. The transmission bandwidth in future 1.55μm optical fiber communications are theoretically given.

II. Structure and Basis of DSM lasers

Typical spectral property of conventional single-mode laser is shown in Fig.1. A single-mode operation is observed in DC bias condition, but such mode hops from one

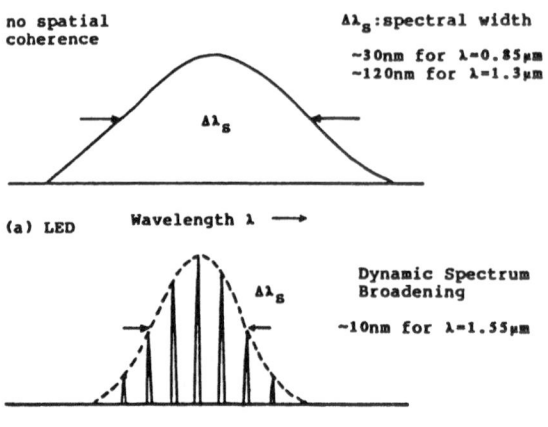

Spectrum in Various Light Sources

no spatial coherence

$\Delta\lambda_s$:spectral width

~30nm for $\lambda=0.85\mu m$
~120nm for $\lambda=1.3\mu m$

$\Delta\lambda_s$

(a) LED   Wavelength $\lambda \longrightarrow$

Dynamic Spectrum Broadening

~10nm for $\lambda=1.55\mu m$

$\Delta\lambda_s$

Wavelength $\lambda \longrightarrow$

(b) Rapidly Modulated Conventional LD

Dynamic Wavelength Shift

$\Delta\lambda_s : 0.1\sim0.3nm$

$\Delta\lambda_s$

Wavelength $\lambda \longrightarrow$

(c) Rapidly Modulated DSM LD

Fig. 1 Conceptional diagrams of the lasing spectrum of a conventional Fabry-Perot type laser (a) under DC operation (b) with the variation of temperature or bias current and (c) under a rapid direct modulation

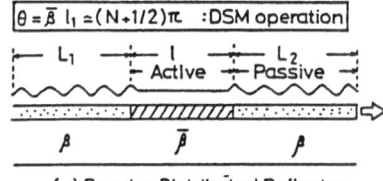

$\boxed{\theta = \bar{\beta}\, l_1 = (N+1/2)\pi \quad :DSM\ operation}$

L_1 — l — L_2
Active — Passive

β $\bar{\beta}$ β

**(a) Passive Distributed Reflector
[DBR Type]**

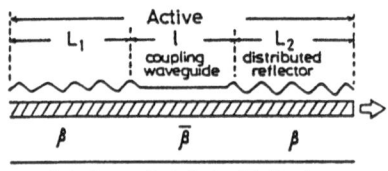

Active

L_1 — l — L_2
coupling waveguide | distributed reflector

β $\bar{\beta}$ β

**(b) Active Distributed Reflector
[DFB Type]**

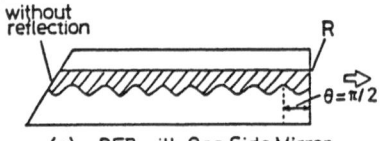

without reflection   R

$\theta=\pi/2$

(c) DFB with One Side Mirror

Fig. 2 Schematic diagram of dynamic-single-mode (DSM) laser with distributed reflectors:
(a) passive distributed reflector laser (DBR type)
(b) active distributed reflector laser (DFB type)
(c) DFB laser with one side facet mirror

mode to another due to the variation of temperature or bias current. Such mode jump also occurs when it is modulated. The lasing spectrum tends to be multimode when the laser is modulated rapidly, as shown in Fig.1c [3],[4]. In contrast, such mode jump or the multimode operation does not occur in a DSM laser.

Figure 2 shows schematic structures of typical distributed reflector type DSM lasers. The amount of phase shift at the central coupling region for the single longitudinal mode operation is essentially same both for the DBR type passive distributed reflector laser shown in Fig.2a and the DFB type active distributed reflector type laser shown in Fig.2b. The required phase shift in the coupling region for the single mode operation is nearly equal to $\pi/2$ relative to the corrugation period, as shown Fig.3. As this phase shift is relative value to the waveguide with grating which is coherent in both DBR regions, the required tolerance is fairly larger.

To avoid two mode behaviors in the axially symmetric structure for DFB laser, the DFB laser with one side mirror, that gives axially antisymmetric structure as shown in Fig.2c, has been tested for the single mode operation. However the mode selectivity of the DFB laser with one side facet mirror is relatively smaller than that of the lasers with phase adjusting region as mentioned in Fig.2a and 2b. The lasing wavelength does approximately fit to the Bragg wavelength, namely twice of the grating pitch multiplied by the effective refractive index.

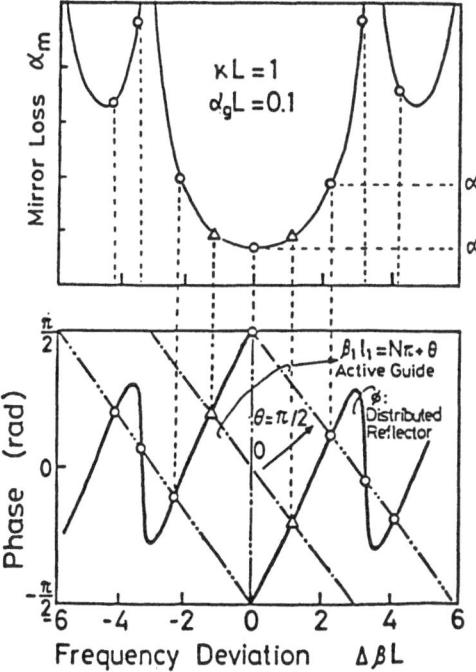

Fig. 3 Schematic diagrams of a threshold gain and a phase shift of DBR as a function of a frequency deviation $\Delta\beta L$ from the Bragg condition. Resonant modes arise at cross points of phase curve for a DBR and that for the active region, as indicated by circles

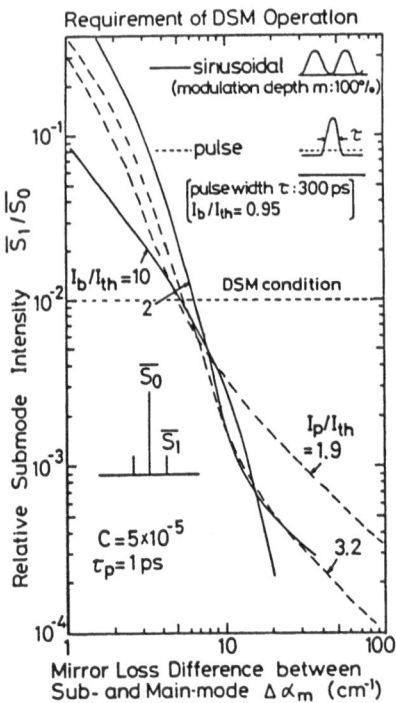

Requirement of DSM Operation

— sinusoidal ⋀⋀
(modulation depth m:100%)

⋯⋯pulse ⋯⋯↗↖τ⋯

[pulse width τ : 300 ps]
[I_b/I_{th}= 0.95]

I_b/I_{th} =10

2

DSM condition

$\overline{S}_0$

$\overline{S}_1$

I_p/I_{th} =1.9

$C=5×10^{-5}$
$τ_p=1ps$

3.2

Mirror Loss Difference between
Sub- and Main-mode $Δα_m$ (cm^{-1})

Relative submode intensity to main mode
under direct modulation

Fig. 4 Normalized submode photon intensity of a rapidly modulated DSM laser as a function of the difference of mirror loss between sub- and main modes at the resonance-like-frequency, where spontaneous emission factor is assumed to be $5×10^{-5}$ [34].

III. Dynamic Single Mode Condition

The dynamic single mode operation is maintained when certain additional resonator loss is imposed to the neighbouring mode around the main lasing mode. Figure 4 shows the relation between the relative neighbouring submode intensity and the loss difference between submode and main mode at modulated conditions of 100 percent sinusoidal modulation with a parameter of the relative bias level I_b/I_{th} and the 300psec pulse modulation at the bias level of $0.95I_{th}$ with a parameter of relative peak current I_p/I_{th} [34],[59]. For example, the loss difference of 5-7cm^{-1} is required to suppress the neighbouring submode less than the main lasing mode by one percent.

Figure 5 shows the loss difference between submode and main mode as a function of additional phase shift at the coupling region or the phase between the top of corrugation shape and the one side mirror[29],[43] for each structure shown in Fig.2. The region for the phase shift required for DFB laser with one side mirror is relatively severe, whereas the region for the additional phase shift required for distributed reflector lasers shown in Fig.2a and b is relatively larger [43]. This additional phase shift is attained by shifting the corrugation pitch at the central area [44],[30] or slightly changing the geometrical structure of the coupling waveguides [45]. The tolerance to satisfy this condition is appreciably larger.

Mirror Loss Difference
in Distributed Reflector DSM Lasers

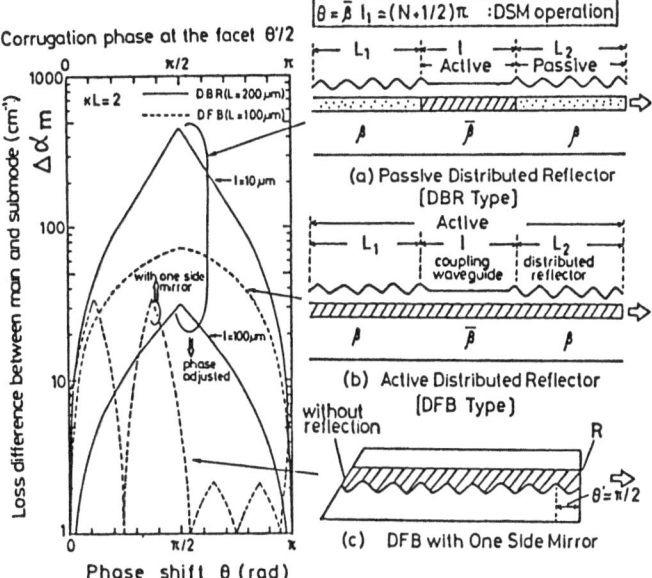

Fig. 5 Loss difference between sub- and main modes as a function of phase shift in the coupling region or the phase difference in the corrugation pitch and one side facet mirror [43].

The effect of the reflection noise in these DSM lasers could be reduced by increasing the reflectivity with increasing the product of the coupling constant of the grating and the grating length [60].

IV. Properties of DSM lasers

As the loss difference between sub- and main modes of typical experimental laser was mostly of the order of several inverse centimeters, therefore the intensity of the submode was observed to be less than one percent and some was less than 0.1 percent compared with the main lasing mode [61]. The mode jump was not observed during the temperature region of 100deg with the wavelength-temperature coefficient about 0.1nm/deg [37], due to the larger area of reflector, higher power handling capability was given with the output power of 60mW at the wavelength of 1.3µm in Fig.6 [48].

The lasing spectrum of a conventional Fabry-Perot single-mode semiconductor laser is turned to be multi-mode or is broadened under the rapid direct modulation as shown in Fig.7b. Lasing mode of those lasers sometimes hops from one to another, even if it operates in single-mode at the DC operation. Thus the spectral mode partition noise occurs as shown in Fig.8, which limits the transmission bit rate in fiber transmission. The dynamic spectral broadening as wide

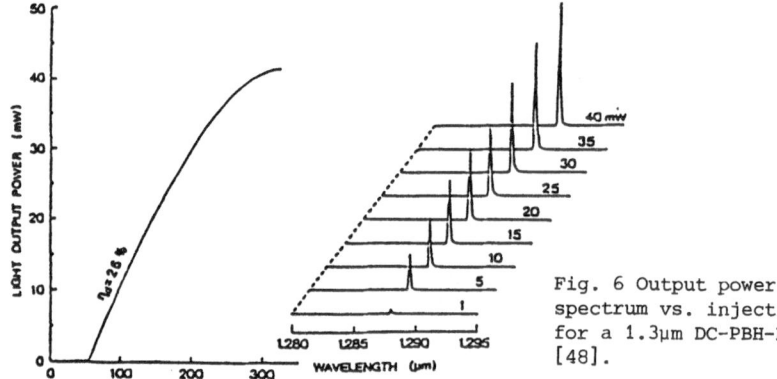

Fig. 6 Output power and lasing spectrum vs. injection current for a 1.3µm DC-PBH-DFB laser [48].

Fig. 7 Time-averaged lasing spectrum under a rapid direct modulation of (a) a BH-DBR-BJB laser in comparison with that of (b) a conventional BH laser [34].

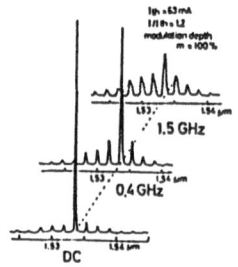

(a) DSM LD   (b) Conventional LD

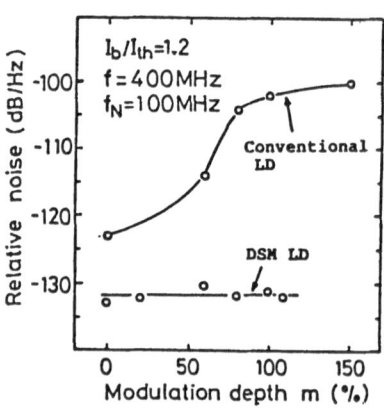

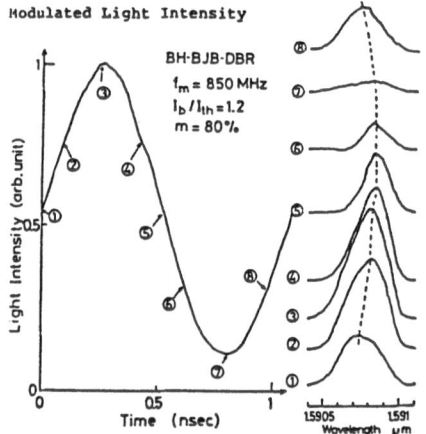

Fig. 8 Relative intensity noise in a lasing mode for conventional and DSM lasers

Fig. 9 Modulated light intensity and corresponding time resolved lasing spectra of a BH-BJB-DBR laser

88

as about 10nm was observed in a conventional 1.5-1.6μm wavelength GaInAsP/InP laser. Whereas the dynamic-single-mode operation is attained in DSM lasers as shown in Fig.7a. However, the lasing wavelength of DSM lasers shifts periodically within a period of modulation, as shown in Fig.9, due to the variation of the refractive index which is originated by oscillation of the carrier density. This phenomenon is called the dynamic wavelength shift [34],[4]. Therefore, the dynamic spectral width of DSM laser is equivalent with the dynamic wavelength shift [34].

The spectra of GaInAsP/InP BH-BJB-ITG laser with the emitting wavelength of 1.6μm under direct rapid modulation were observed by use of the monochromator, where the modulation depth was 100 percent and the bias current level I_b of 1.2 times the threshold. The modulation frequency was increased from 0.5GHz to 1.9GHz. The time-averaged lasing spectrum under rapid direct modulation of a DSM laser is compared with that of the conventional laser as shown in Fig.7a. A stable dynamic-single-mode operation was observed all over the modulation frequencies as shown in this figure. We note that the line width of the lasing mode was broadened due to the dynamic wavelength shift and it became maximum at the resonance-like frequency of 1.9GHz. The dynamic wavelength shift of the BH-DBR-ITG laser increased appreciably when the modulation depth m increased over than 50 percent at the frequency of 1.9GHz. The full width at half maximum intensity of the spectrum without modulation was observed here to be about 0.07nm, which was the resolution limit of monochromator. Subtracting this spectral resolution from the line width under the direct modulation, the full width of the dynamic wavelength shift $\Delta\lambda_{s,max}$ was 0.27nm at the resonance-like frequency of 1.9GHz with m=100% and it was about 1/37 times the dynamic spectral width of conventional lasers emitting at same wavelength [47].

The modulation-frequency dependence of the dynamic wavelength shift numerically calculated using rate equations was in good agreement with experiment [47].

The extreme value of the dynamic line shift is determined solely by the ratio of the imaginary and real parts of refractive variation of the active region [62].

The reliability test has been undertaken, and there is no indication that reliability of DSM lasers is inferior to that of a conventional laser [53],[54]. As the problem at the mirror facet for the case of a conventional laser is eliminated for DSM lasers, the operation lifetime could be significantly increased by reducing the threshold current density.

The modulation sensitivity is essentially same to that of conventional lasers, therefore the resonance-like peak appearing at the sinusoidal modulation or the relaxation oscillation appearing at the pulse modulation can be minimized by narrowing the stripe width down to the carrier diffusion-length of a few micrometers. As it is possible to increase the mirror reflectivity of DSM lasers, the problem of reflection noise could be reduced further. The partition noise generated by the mode jump was observed to be eliminated, as shown in Fig.8 [63].

V. Active Semiconductor Integrated Optics

The monolithic integration of various optical devices and electronic devices are underway for better performances of laser diodes and optical devices.

For active semicondutor integrated optics with possible multi-purpose use, integration is used to improve optical devices such as lasers [64], and to integrate both optical and electronic devices so as to achieve better matching between optical devices and electronic circuits [65]-[69]. Since the mobilities of the substrate materials, GaAs and InP, are large, the possibility of integrating high speed electronic devices together with optical devices is promising. Integration of a laser with a transistor can increase the input impedance for laser modulation, and integrating photodiode together with an FET can decrease the output impedance [65]-[67].

Development of integrated lasers with the output waveguide will play a central role in integrated optics and are of particular interest in semiconductor monolithic integrated circuits [64]. Experiments on the monolithic integration of lasers with light amplifiers, modulators, and detectors have been reported [70]-[76]. The study of integrated optics is thus keeping in step with the development of optical electronics.

VI. Application to Single Mode Fiber Communication

Very typical future application of DSM lasers is the wideband optical fiber transmission in the loss-minimum wavelength of 1.5 to 1.6μm. The transmission bandwidth of single-mode (SM) fiber is limited by the chromatic dispersion τ_c [57]. The transmission bandwidth B of a conventional single-mode fiber of the length L with the chromatic dispersion τ_c per unit length and unit spectral width of light source is [2],

$$B = 1/(\tau_c \Delta\lambda_s L) \tag{1}$$

where $\Delta\lambda_s$ is the spectral width of light source assumed to be larger than that of corresponding modulation frequency. When the spectral width of the light source is smaller than that of the corresponding modulation bandwidth $\Delta\lambda_{mod}$, the transmission bandwidth B is [8],

$$B = 1/(\tau_c \Delta\lambda_{mod} L) = \{c/(2\lambda^2 \tau_c L)\}^{1/2} \tag{2}$$

This equation gives the theoretical limit of the transmission bandwidth of a single-mode fiber. However, there is some optimum condition for the pulse width, and the extreme relative bandwidth can be shown to be [62],

$$BL^{1/2} = 1/[2\{2(1+\alpha^2)^{1/2} + 2\alpha\}^{1/2}(|\ddot{\beta}L|)^{1/2}]$$
$$= 25 \text{ Gbits/s(km)}^{1/2} \text{ (for } \alpha=4) \tag{3}$$

The fiber transmission experiments at 1.55μm wavelength region are summarized in Table 1. Some experiment shows that

Table1   Transmission experiments with 1.55μm DSM lasers.

$$\left(\begin{array}{l} \text{Theoretical} \\ \text{Dispersion Limit } B \cdot \sqrt{L} = \sim 25 Gbit/s \cdot km^{1/2} \end{array} \right)$$

the experimental results already reached to the theoretical
limitations [56].   As shown in eq.(3), the bandwidth is
theoretically dependent on the factor α, the ratio of real
part to imaginary part of carrier induced refractive index

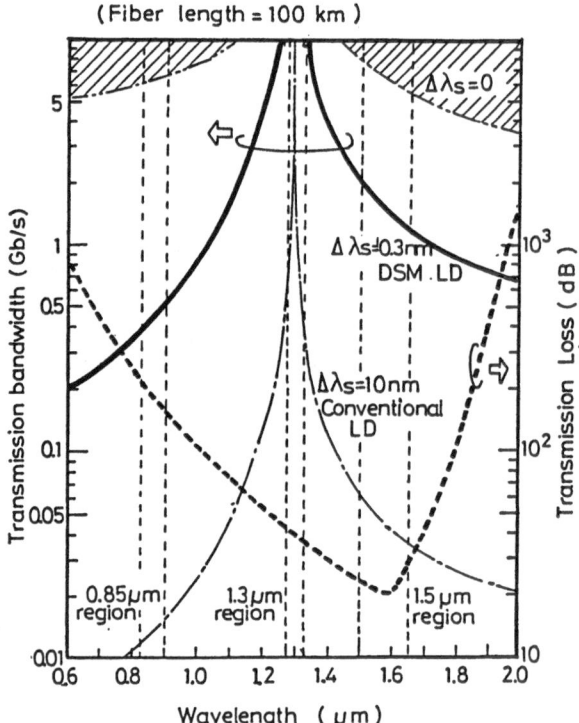

Fig. 10 Transmission bandwidth B of a conventional single-mode fiber of 100km long as a function of the wavelength of light source with a parameter of the dynamic spectral width of it. Dashed thick line indicates the transmission loss characteristics of the single-mode fiber [62]

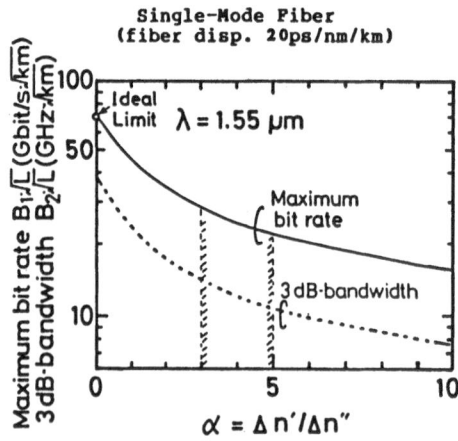

Fig. 11 Relation between transmission bandwidth of a single-mode fiber and the ratio of real and imaginary parts of carrier injection induced refractive index change, α, of a DSM laser

change. This condition is shown in Fig.10. Theoretical limit can be increased by reducing α.

Figure 11 shows the transmission bandwidth B of a conventional single-mode fiber of 100km long as a function of the wavelength of light source. Solid lines were for transmission bandwidth and the dotted line gives the fiber transmission loss.

From above discussion, introduction of a DSM laser such as DBR and DFB lasers will enable us to use high capacity optical fiber communications at the loss-minimum-wavelength of 1.5 to 1.6µm. Furthermore, the mode partition noise generated by lasing mode-jump during high frequency modulation [77],[78] will be significantly reduced by use of the DSM lasers.

These DSM lasers with expected significant features against the noise due to the reflected light wave, which can be realized by using higher reflectivity of DBR section, and also against the mode hopping could be applied for other fields such as the optical video disk, the fiber gyroscope [79] and the optical measurements.

REFERENCES

1   T.Miya, Y.Terunuma, T.Hosaka and T.Miyashita,''An ultimately low-loss single-mode fiber at 1.5µm,'' Electron. Lett., vol.15,No.4,pp.106-108, Feb. 1979.

2   W.A.Gambling, H.Matsumura, and C.M.Ragdale, ''Total dispersion in graded index single-mode fibers', Electron. Lett., vol. 15, no. 15, pp.474-476, 1979.

3   T.Ikegami,''Spectrum broadening and tailing effect in direct modulated injection lasers,'' Proc. 1st European Conf. Optical Fiber Commun., London, pp.111, 1975.

4   K.Kishino, S.Aoki and Y.Suematsu, ''Wavelength variation of 1.6µm wavelength buried heterostructure GaInAsP/InP lasers due to direct modulation,'' IEEE J. Quantum Electron., vol.QE-18, no.3, pp.343-351, Mar. 1982.

5   K.Utaka, K.Kobayashi, K.Kishino and Y.Suematsu, ''1.5-1.6µm GaInAsP/InP integrated twin-guide lasers with first-order distributed Bragg reflectors,'' Electron. Lett., vol.16. No.12, pp.455-456, June 1980.

6   K.Utaka, K.Kobayasi, and Y.Suematsu, 'Lasing characteristics of 1.5-1.6µm GaInAsP/InP integrated twin-guide lasers with distributed Bragg reflectors', IEEE J. Quantum Electron., vol. QE-17, No.5, pp.651-658, May 1981.

7   K.Utaka, Y.Suematsu, K.Kobayashi and H.Kawanishi, ''GaInAsP/InP integrated twin-guide lasers with first order distributed Bragg reflector at 1.3µm wavelength'', Japan J. Appl. Phys.,vol.19, pp.137-140, 1980.

8   Y.Sakakibara, K.Furuya, K.Utaka and Y.Suematsu, ''Single-mode oscillation under high-sped direct modulation in GaInAsP/InP integrated twinguide lasers with distributed-Bragg-reflectors,'' Electron. Lett., vol.16, No.12,pp.456-458, June 1980.

9   O.Mikami, ''1.55µm GaInAsP/InP distributed feedback lasers'', Japan. J. Appl. Phys., vol.20, No.7, pp.L488-L490, July 1981.

10   K.Utaka, S.Akiba, K.Sakai and M.Matsushima, ''Room-temperature CW operation of distributed-feedback buried-heterostructure InGaAsP/InP lasers emitting at

1.57μm," Electron.Lett., vol.17, No.25/26, pp.961-963, Dec.1981.

11 T.Matsuoka, H.Nagai, Y.Itaya, Y.Noguchi, U.Suzuki and T.Ikegami, "CW operation of DFB-BH GaInAsP/InP lasers in 1.5μm wavelength region," Electron. Lett., vol.18, No.1, pp.27-28, Jan. 1982.

12 Y.Abe, K.Kishino, Y.Suematsu and S.Arai,"GaInAsP/InP integrated laser with butt-jointed distributed-Bragg-reflector waveguide," Electron. Lett., vol.17, No.25/26, pp.945-947, Dec.1981.

13 Y.Suematsu, S.Arai and K.Kishino,"Dynamic single-mode semiconductor lasers with a distributed reflector", IEEE J..Lightwave Tech., vol.LT-1, No.1, pp.161-176, Mar.1983.

14 H.Kogelnik and C.V.Shank,"Stimulated emission in a periodic structure," Appl.Phys.Lett., vol.18, pp.152-154, 1971.

15 H.Kogelnik and C.V.Shank, "Coupled-wave theory of distributed feedback lasers," J.Appl.Phys., vol.43, No.5, pp.2327-2335, May 1972.

16 M.Nakamura, A.Yariv, H.W.Yen, S.Smoke and H.L.Garrin," Optically pumped GaAs surface laser with corrugation feedback," Appl. Phys. Lett., vol.22, pp.515-516, 1973.

17 C.V.Shank, R.V.Schmidt, and B.I.Miller, "Double-heterostructure GaAs distributed-feed back laser," Appl. Phys. Lett., vol.25, No.4, pp.200-201, Aug. 1974.

18 D.R.Scifres, R.D.Burnham, and W.Streifer, "Distributed-feedback single heterojunction GaAs diode laser," Appl. Phys. Lett, vol.25, No.4, pp.203-206, Aug.1974.

19 H.M.Stoll and D.H.Seib, "Distributed feedback GaAs homojunction injection laser", Appl. Opt., vol.13, No.9, pp.1981-1982, Sept. 1974.

20 M.Nakamura, K.Aiki, J.Umeda, A.Yariv, H.W.Yen, and T.Morikawa, "GaAs-Ga$_{1-x}$Al$_x$As double heterostructure distributed feedback diode lasers," Appl. Phys. Lett., vol.25, No.9, pp.487-488, Nov.1974.

21 D.B.Anderson, R.R.August, and J.E.Coker, "Distributed-feedback double-heterostructure GaAs injection laser with fundamental grating," Appl.Opt., vol.13, No.12, pp.2742-2744, Dec.1974.

22 K.Ishii, "Emission properties of AlGaAs semiconductor lasers below and above threshold," Bachelor Thesis, Tokyo Institute of Technology, 1974.

23 Y.Suematsu, M.Yamada and H.Hayashi,"A multi-hetero AlGaAs laser with integrated twin guide", Proc. IEEE, vol.63, p.208, 1975.

24 I.P.Kaminow and H.P.Weber, "Poly (Methyl Methacrylate) Dye laser with internal diffraction grating resonator," Appl. Phys. Lett. vol.18, pp.497-499, 1971.

25 F.K.Reinhart, R.A.Logan, and C.V.Shank, "GaAs-Al$_x$Ga$_{1-x}$As injection lasers with distributed Bragg reflectors," Appl. Phys. Lett., vol.27, No.1, pp.45-48, July 1975.

26 W.T.Tsang and S.Wang, "GaAs-Ga$_{1-x}$Al$_x$As double-heterostructure injection lasers with distributed Bragg

reflectors,'' Appl. Phys. Lett., vol.28, No.10, pp.596-598, May 1976.

27   N.Matsumoto and K.Kumabe, ''Influence of structure parameters on the lasing characteristics of semiconductor lasers,'' Japan. J. Appl. Phys., vol.18, No.2, pp.321-332, Feb. 1979.

28   T.Kuroda, S.Yamanishi, M.Nakamura and J.Umeda, ''Channeled-substrate-planar structure distributed feedback semiconductor lasers,'' Appl. Phys. Lett., vol.33, no.2, pp.173-174, July 1978.

29   W.Streifer, R.D.Burnham and D.R.Scifres, ''Effect of external reflectors on longitudinal modes of distributed feedback lasers,'' J. Quantum Electron., vol.QE-11, No.4, pp.154-161, Apr. 1975.

30   H.A.Haus and C.V.Shank, ''Antisymmetric taper of distributed feedback lasers,'' IEEE J. Quantum Electron., vol.QE-12, No.9, 1976.

31   A.Suzuki and K.Tada, ''Theory and experiment on distributed feedback lasers with chirped grating,'' Proc. of SPIE, vol.239, Guided-Wave Optical and Surface Accoustic Wave Devices, Systems and Applications, San Diego, pp.10-18, 1980.

32   W.T.Tsang, N.A.Olsson, R.A.Linke, R.A.Logan ''1.5μm wavelength GaInAsP C^3 lasers single-frequency operation and wide band frequency tuning,'' Electron. Lett., vol.19, No.11, pp.415-416, May 1983.

33   K.Iga and Y.Takahashi, ''An analysis on single wavelength oscillation of semiconductor lasers of high speed direct pulse modulation,'' Trans. IECE Japan, vol.E-61, no.9, pp.685-689, Sept. 1978.

34   F.Koyama, Y.Suematsu, S.Arai, and T.Tanbun-ek, ''1.5-1.6μm GaInAsP/InP dynamic-single-mode (DSM) lasers with distributed Bragg reflector,'' IEEE J. Quantum Electron., vol.QE-19, No.6, June 1983.

35   Y.Itaya, T.Matsuoka, K.Kuroiwa and T.Ikegami,''Longitudinal mode spectra of 1.5μm GaInAsP/InP distributed feedback lasers'', 4th Integrated Optics and Optical Fiber Comm. Conf. (IOOC), Tokyo, 28B1-1,1983.

36   S.Akiba, K.Utaka, K.Sakai and Y.Matsushima,''Effect of mirror facet on lasing characteristics of InGaAsP/InP DFB lasers'', ibid, 28B1-2, 1983.

37   S.Akiba, K.Utaka, K.Sakai and Y.Matsushima,''Low-Threshold-Current Distributed-Feedback InGaAsP/InP CW Lasers'', Electron.Lett., vol.18,No.2, pp.77-78, 1982.

38   Y.Itaya, T.Matsuoka, Y.Nakano, Y.Suzuki, K.Kuroiwa and T.Ikegami, ''New 1.5μm Wavelength GaInAsP/InP Distributed Feedback Laser'', Electron.Lett., vol.18, No.23, pp.1006-1007, 1982.

39   L.D.Westbrook, A.W.Nelson, P.J.Fiddyment and J.S.Evans, ''Continuous-Wave Operation of 15, May 1972.

40   H.Katsuda, T.Watanabe, S.Suzaki, T.Takahashi, T.Nomura, T.Shimme, 'Oscillation characteristics of distributed feed back laser in 1.5μm wavelength region', Nat.Conv.Rec. of IECE Jpn. 1020, March, 1984.

41   K.Fujiwara, M.Morimoto, T.Tanahashi, H.Ishikawa, H.Imai, '1.55μm DFB-BH laser', Nat.Conv.Rec. of Jpn.Appl.Phys., 1a-M-10, March, 1984.

42 T.L.Koch, T.J.Bridges, E.G.Burkhardt, R.A.Logan, L.F.Johnson, L.A.Coldren, P.J.Corvini, and W.t.Tsang, ''1.55μm InGaAsP vapor phase transported buried heterostructure distributed feedback lasers,'' 9th IEEE Int. Semiconductor Laser Conf., Post deadline paper No.6, Rio de Janeiro, Aug. 1984.

43 N.Eda, K.Sekartedjo, K.Furuya, and Y.Suematsu, 'Dynamic single mode characteristics of phase shifted DFB laser', Nat.Conv.Rec. of IECE Jpn. 984, March, 1984.

44 K.Sekartedjo, N.Eda, K.Furuya, Y.Suematsu, F.Koyama, and T.Tanbun-ek,'' 1.5μm phase-shifted DFB lasers for single-mode operation ,'' Electron. Lett. vol.20, pp.80-81, 1984.

45 F.Koyama, Y.Suematsu, K.Kojima and K.Furuya, ''1.5μm phase adjusted active distributed reflector laser for complete dynamic single mode operation'', to be appeared in Electron.Lett.

46 Y.Itaya, K.Wakita, G.Motosugi, and T.Ikegami, ''Phase control by coating in 1.5μm DFB lasers,'' 9th IEEE Int. Semiconductor Laser Conf., E-1, Rio de Janeiro, Aug. 1984.

47 F.Koyama, S.Arai, Y.Suematsu, and K.Kishino, 'Dynamic spectral width of rapidly modulated 1.58μm GaInAsP/InP buried-heterostructure distributed-Bragg-reflector integrated-twin-guide lasers,' Electron. Lett., vol.17, no.25/26, pp.938-940, Dec. 1981

48 M.Kitamura, M.Seki, M.Yamaguchi, I.Mito, Ke.Kobayashi, and Ko.Kobayashi,''High-Power Single-Longitudinal-Mode Operation of 1.3μm DFB-DC-PBH LD,'' Electron.Lett., Vol.19, No.20, pp.840-841, 1983.

49 Y.Suzuki, H.Nagai, Y.Noguchi, T.Matsuoka, and K.Kurumada, ''High power SLM operation of 1.3μm InP/GaInAsP DFB LD with doubly buried heterostructure on p-type InP substrate,'' Electron. Lett., vol.20, No.21, pp.881-882, Oct. 1984.

50 H.Okuda, Y.Hirayama, J.Kinoshita, H.Furuyama and Y.Uematsu, ''High-Quality 1.3μm GaInAsP/InP BH-DFB Lasers with First-Order Gratings'', Electron.Lett., vol.19,No.22,pp.941-943, 1983.

51 H.Asahi, Y.Kawamura, Y.Noguchi, T.Matsuoka and H.Nagai, ''Hybrid LPE/MBE-Grown InGaAsP/InP DBR Lasers'', Electron.Lett., vol.19,No.14, pp.507-508, 1983.

52 B.Broberg, F.Koyama, Y.Tohmori, and Y.Suematsu, ''1.53μm DFB lasers by mass transport,'' Electron. Lett., vol.20, No.17, pp.692-694, Aug. 1984.

53 G.Motosugi, Y.Nakao, Y.Yoshikuni, and T.Ikegami, 'Aging characteristics of InGaAsP/InP DFB lasers', 9th European Conference on Optical Commun., pp.43-46, Geneva Oct. 1983.

54 S.Akiba, K.Utaka, K.Sakai, and Y.Matsushima, 'Aging test of 1.5μm InGaAsP/InP DFB laser', Nat.Conv.Rec. of Jpn. Appl. Phys., 1p-M-10, Apr. 1984.

55 T.Ikegami, K.Kuroiwa, Y.Itaya, S.Shinohara, K.Hagimoto, and N.Inagaki, ''1.5μm transmission experiment with distributed feedback laser,'' 8th European Conf. Opt. Fiber Commun., Cannes, 1982.

56 B.L.Kasper, R.A.Linke, K.L.Walker, L.G.Cohen, T.L.Koch, T.J.Bridges, E.G.Burkhardt, R.A.Logan, R.W.Dawson, and J.C.Campbell, 'A 130km transmission experiment at 2Gb/s using silica-core fiber and a vapor phase transported DFB laser', ECOC'84 post deadline papers, PD-6, Sep. 1984

57 Y.Ichihashi, H.Nagai, T.Miya, and Y.Miyajima, ''Transmission experiment over 134km of single-mode fiber at 445.8Mb/s,'', IOOC'83, Post deadline paper 29C5-2, Tokyo, July 1983.

58 V.J.Mazurczyk, N.S.Bergano, R.E.Wagner, K.L.Walker, N.A.Olsson, L.G.Cohen, R.A.Logan, and J.C.Campbell, ''420Mb/s transmission through 203km using sillica core fiber and a DFB laser,'' ECOC'84, Post deadline paper PD-7, Sept.1984.

59 S.Takahashi, H.Tsushima, F.Koyama, and Y.Suematsu, 'Single mode condition of pulsed modulated semiconductor lasers', Nat.Conv.Rec. of IECE Jpn. 981, March, 1984.

60 F.Koyama, and Y.Suematsu, 'Analysis of reflection noise in dynamic single mode lasers', Nat.Conv.Rec. of Jpn.Appl.Phys., 29p-M-11, March, 1984.

61 T.Tanbun-ek, S.Suzaki, F.Koyama, S.Arai and Y.Suematsu, ''Static lasing characteristic of CW operation of 1.5-1.6μm GaInAsP/InP''

62 F.Koyama, and Y.Suematsu, 'Dynamic wavelength shift of dynamic single mode (DSM) lasers and its influence on the transmission bandwidth of single mode fibers', Tech. Group Meeting of IECE Jpn. OQE 84-71 pp.23-30 July 1984.

63 F.Koyama, T.Tanbun-ek, S.Arai, S.Wang, Y.Suematsu, and K.Furuya, 'Suppression of intensity fluctuation of a longitudinal mode in directly modulated GaInAsP/InP dynamic-single-mode laser', Electron. Lett. vol.19, no.9, pp325-327, Apr. 1983.

64 S.Kawakami and J.Nishizawa, ''An optical waveguide with optimum distribution of the refractive index with reference to waveform distortion,'' IEEE Trans. Microwave Theory and Techniques, vol.MTT-16, No.10, pp.814-818, Oct. 1968.

65 D.Marcuse, ''Losses and impulse response of a parabolic index fiber with random bends,'' Bell Syst. Tech. J., vol.52, pp.1423-1437, Oct. 1973.

66 R.Olshansky and D.B.Keck, ''Pulse broadening in graded-index optical fibers,'' Appl. Opt., vol.15, pp.483-491, Feb. 1976.

67 M.Eve, ''Multipath time dispersion of an optical network,'' Opt. Quantum Electron., vol.10, pp.41-51, Jan. 1978.

68 M.Nakahara, S.Sudo, N.Inagaki, K.Yoshida, S.Shibata, K.Kokura, and T.Kuroha, ''Ultra wide bandwidth VAD Fiber,'' Electron. Lett., vol.16, No.10, pp.391-392, May 1980.

69 K.Okamoto, T.Edahiro, and M.Nakahara, ''Transmission characteristics of VAD multimode optical fibers,'' Appl. Opt., vol.20, No.13, pp.2314-2318, July 1981.

70 K.Ogura, S.Shibutani, K.Yoshida, S.Sudo, and M.Nakahara, ''Fabrication of side-band VAD optical

fiber," National Convention on Optical and Ratio Wave Section, Inst. Electron. Comm. Engrg. Japan, 316, Oct. 1980.

71 A.J.Ritger and F.T.Stone, "Improving the bandwidth of optical fibers made using modified CVD," Topical Meeting on Opt. Fiber Commun., Phoenix, TUEE4, Apr. 1982.

72 F.Kapron, "Maximum information capacity of fiber-optic waveguide," Electron. Lett., vol.13, pp.96-97, Feb. 1977.

73 W.A.Gambling and H.Matsumura, "Propagaion in radially-homogeneous single-mode fibere," Opt. Quantum Electron., vol.10, pp.31-40, Jan. 1978.

74 A.W.Snyder and R.A.Sammut, "Fundamental (HE_{11}) modes of graded optical fibers," J. Opt. Soc. Amer., vol.69, pp.1663-1671, Dec. 1979.

75 H.Tsuchiya and N.Imoto, "Single mode fiber delay equalization and baseband response," IECE Japan Eng. Res. Rep. OQE 79-22, May 1979.

76 L.G.Cohen and C.Lin, "Pulse delay measurements in the zero material dispersion wavelength region for optical fibers," Appl. Opt., vol.12, pp.3136-3139, Dec. 1977.

77 Y.Okano, K.Nakagawa, and T.Ito, 'Laser mode partition noise evaluation for optical fiber transmission', IEEE Trans. Commun., vol.com-28, no.2, pp.238-243, Feb. 1980.

78 K.Ogawa, and R.S.Vodhanel, 'Analysis and measurement of mode partition noise,' Topical Meeting on Opt.Fiber Commun., Phoenix, THDD4, Apr. 1982

79 R.A.Bergh, C.C.Cutler, H.C.Lefevre, S.A.Newton, G.A.Pavlath, and H.J.Shaw, 'The all fiber gyroscope: A practical alternative for rotation sensing,' Topical Meeting on Integrated and Guided Wave Optics, pacific Grove, WB-2, Jan. 1982.

Optical Waveguide Properties of Multiquantum Wells

M. Glick and F.-K. Reinhart

Institut de Physique appliquée, Ecole Polytechnique Fédérale de Lausanne
PH-Ecublens, CH-1015 Lausanne, Switzerland

We have studied the optical waveguide properties of a GaAs-AlGaAs multi-quantum well waveguide and determined the linear electro-optic coefficient, r_{41} = -1.47 x 10^{-10} cm/V at λ = 1.1523 µm.

Recent research has generated considerable interest in the special properties of multiquantum well (MQW) structures. But although their values as diode lasers[1] and modulators[2] have been studied, little is known concerning their properties as optical waveguides. We are currently studying these properties, and in particular we are examining the electro-optic effects. We have determined the linear electro-optic (LEO or Pockel's) effect in a GaAs-AlGaAs MQW heterostructure p-n junction diode at λ = 1.1523 µm.

Our initial study, far from the observed room-temperature excitonic peak of 0.858 µm, is at λ = 1.1523 µm where the LEO effect dominates. We are continuing our measurements at other wavelenghts to complement our value for the LEO coefficient, r_{41}, and to examine higher order effects, as the quadratic effect has never been determined in a GaAs heterostructure. Recent studies in InGaAsP/InP double heterostructures[3] have identified a quadratic electro-optic effect strongly dependent on the wavelength and on the bandgap. Preliminary measurements lead us to believe that similar results may be found as we approach the excitonic peak of the MQW structure. The layering of the MQW structure gives a relatively high birefringence[4]. The layering may also add other morphological effects[5] modifying the electro-optic effect.

A schematic of the MQW structure used in our experiment is shown in Fig. 1. For initial measurements we used a sample of area 1.09 x 0.906 mm^2 cleaved to allow measurements of the phase modulation along the orthogonal crystallographic [110] and [$\bar{1}$10] directions.

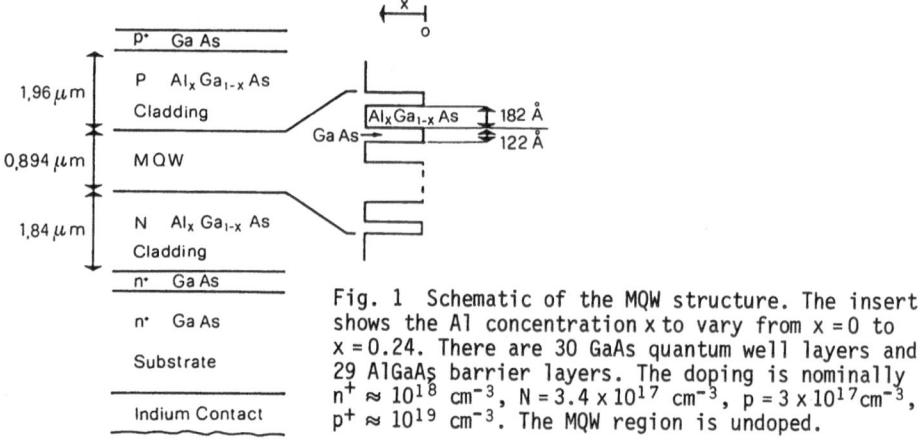

Fig. 1 Schematic of the MQW structure. The insert shows the Al concentration x to vary from x = 0 to x = 0.24. There are 30 GaAs quantum well layers and 29 AlGaAs barrier layers. The doping is nominally $n^+ \approx 10^{18}$ cm^{-3}, $N = 3.4 \times 10^{17}$ cm^{-3}, $p = 3 \times 10^{17}$ cm^{-3}, $p^+ \approx 10^{19}$ cm^{-3}. The MQW region is undoped.

A helium-neon laser operating at λ = 1.1523 μm was the source for the initial measurements. To study the dispersion of the phase modulation and complement initial measurements, we are using an argon-ion laser operating at λ = 1.0923 μm, for wavelengths in the region of the observed excitonic peak we use the argon-ion laser to pump a dye laser using Styryl 9 dye.

Our MQW waveguide was found to have extremely low loss, causing strong Fabry-Perot resonances, which made it difficult to measure the waveguide loss and phase modulation. We found it necessary to provide the four mirror faces with anti-reflection (AR) coatings to suppress the resonances. We formed the AR-coatings by anodizing the sample in $H_2O + H_3PO_4$ at pH = 2.5 to 82V. With these conditions we achieved with an unfocussed beam of λ = 1.1523 μm a reflectivity of 0.9% from a polished GaAs surface. We present results from a systematic study of this procedure for various wavelengths and voltages.

The measurements for the phase difference for λ =1.1523 μm are reduced per unit length and shown in Fig. 2. The LEO is clearly dominant. The intersection point gives the built-in phase difference $\Delta\Phi_o$ = 560 degree/mm. The phase differences are -8.31 degree/Vmm and 7.70 degree/Vmm for the [110] and [$\bar{1}$10] directions respectively. For materials of the $\bar{4}$3m group we would expect these values to be equal if only the LEO effect were involved[5]. The measurements yield values of r_{41} = -1.53 x 10^{-10} cm/V and r_{41} = -1.42 x 10^{-10} cm/V associated with the light propagation along the [110] and [$\bar{1}$10] directions respectively. They differ within our estimated error limit of 1.7%. The average value yields r_{41} = [-1.47 ± 0.03] x 10^{-10} cm/V. We note that this value is close to pure GaAs

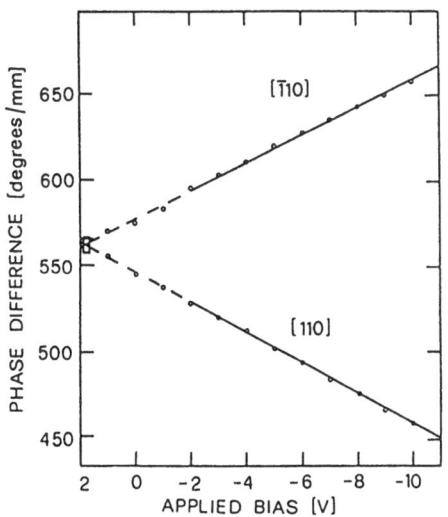

Fig. 2 Phase difference between TE and TM modes as a function of the applied bias for the light propagation along the orthogonal [110] and [$\bar{1}$10] directions. The solid lines are linear fits to the data for the applied bias V_a <- 2V at which point the electric field sweeps out the entire MQW junction regions. The dashed lines are a linear extrapolation for V_a >- 2V.

r_{41} = -1.50 x 10^{-10} cm/V[6]. The estimated value of a homogeneous layer with comparable Al content, $Al_{0.14}Ga_{0.86}As$, is r_{41} = -1.43 x 10^{-10} cm/V.

The discrepancy in the results for the two directions may arise from higher order effects. The small discrepancy implies that if there is a morphic effect it is very weak in this structure. Measurements for InGaAsP/InP double heterostructure diodes show a discrepancy from the LEO effect arising from the quadratic effect[3]. Our deviation from the average linear effect is 3.7% at V_a = -10V. Using this we can estimate the quadratic effect, $R_{11} - R_{12}$ = -5.2 x 10^{-17} cm^2/V. Although we cannot assert a value for the quadratic effect at this time, we expect to present definitive results on the basis of the now ongoing experiments.

Theoretical considerations[8] and previous experiments indicate that the LEO effect and a possible morphic effect should have a relatively weak dispersion,while the magnitude of the quadratic effect should increase significantly as we approach the bandedge.

In conclusion,our measurements have demonstrated that the MQW structure has loss low enough to be suitable for waveguides in connection with other integrated optic devices. Using the MQW as a waveguide,we have determined the LEO coefficient at λ = 1.1523 μm.measurements using other wavelengths are currently in progress to study the dispersion of the electro-optic effects.

References

[1] W.T. Tsang, Appl. Phys. Lett. 39, 786 (1981).

[2] T.H. Wood, C.A. Bunus, D.A.B. Miller, D.S. Chemla, T.C. Damen, A.C. Gossard and W. Wiegmann, Appl. Phys. Lett. 44, 16 (1984).

[3] H.G. Bach, J. Krauser, H.P. Nolting, R.A. Logan and F.K. Reinhart, Appl. Phys. Lett. 42, 692 (1983).

[4] M. Born and E. Wolf, Principles of Optics, 3rd edition (Pergamon Press, Oxford 1965).

[5] J.F. Nye, Physical Properties of Crystals (Clarendon Press, Oxford 1957).

[6] M. Glick, F.K. Reinhart and G. Weimann, Helvetica Physica Acta, accepted for publication 1985.

[7] I.P. Kaminow and E.H. Turner Handbook of Lasers (The Chemical Rubber Co, Cleveland, Ohio 1971), p. 447.

[8] D.B. Kushev, V.I. Sokolov and V.K. Subashier, Sov. Solid State Phys. 13, 2488 (1972).

Derivation of Modelling Parameters for GaInAs(P) Optical Amplifiers and Lasers

S. Hausser, E. Zielinski, H. Eisele, H. Schweizer, and M.H. Pilkuhn

Universität Stuttgart, Physikalisches Institut, Pfaffenwaldring 57
D-7000 Stuttgart 80, Fed. Rep. of Germany

M. Rosenzweig

Heinrich Hertz Institut, Einsteinufer 37, D-1000 Berlin 10, Germany

A method for finding modelling parameters for optical amplifiers and lasers by means of optical gain measurements based on an excitation length procedure is presented.

I. Introduction

To optimize active devices for integrated optical systems it is very important to know the optical gain coefficients as a function of pump intensity, temperature, device length and material composition. The spectral range of 1.3 μm - 1.55 μm is very suited for these devices because of low losses and dispersion in optical fibers. The quaternary and ternary compound semiconductors GaInAs(P) have an emission wavelength between 1.3 and 1.65 μm and are therefore of great interest for active optical devices. However, the optical amplifiers (OA) made from these materials have not been studied experimentally up to now. Experimental investigations are mainly done in the GaAs system /1/. Calculations based on assumptions of optical gain, neglecting gain saturation effects /2/, are done for the 1.55 μm material system /3/. In this paper, we present measurements of the optical gain coefficients by stripe length variation experiments /4/, as a function of pump intensity, temperature and excited stripe length. Our data indicate that the amplifier saturates at a constant product of optical gain times device length, thus limiting the total amplifier gain. Furthermore, based on our data, we have evidence that the T_0-problem /5/ in the 1.6 μm materials is caused by the same microscopic mechanisms as found in the 1.3 μm materials /6/.

II. Experimental

For our measurements, we used conventionally grown LPE layers /7/ lattice matched to InP of high quality. The residual impurity densities range from $8 \cdot 10^{14} \mathrm{cm}^{-3}$ for GaInAs to $2 \cdot 10^{16} \mathrm{cm}^{-3}$ for GaInAsP (λ=1.55μm). Optical excitation of the samples was obtained by a nitrogen laser pumped pulsed dye laser (ν_{rep} = 100 Hz, τ_{pulse} = 10 ns) or a Q-switched Nd:YAG laser (ν_{rep} = 1kHz, τ_{pulse} = 250 ns), focussed onto the samples by a cylindrical lens. Thereby we produced an excited stripe of variable length (L) on the surface of the samples, ranging from 50 to 200 μm, as depicted in fig. 1. The emitted light was dispersed by a high-resolution monochromator and detected by a N_2 cooled Ge-detector. At sufficient pump intensities, amplified spontaneous emission (ASE) parallel to the stripe direction is obtained. Optical gain (g) is computed comparing two ASE spectra of different stripe lengths /4/, interpreting the light output with a one-dimensional amplification geometry:

$$I_{ASE} \sim \frac{r_{spon}}{g} \left(e^{g \cdot L} - 1 \right) \qquad (1)$$

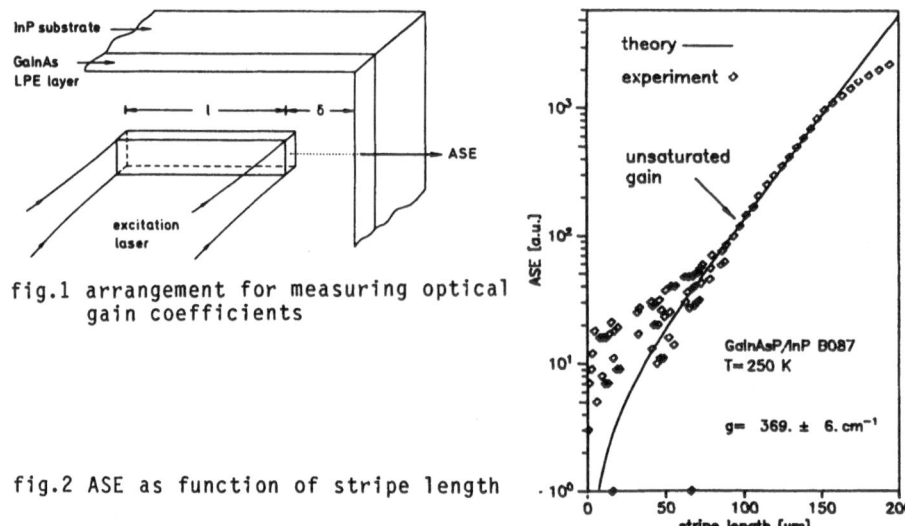

fig.1 arrangement for measuring optical
      gain coefficients

fig.2 ASE as function of stripe length

The measured light output as a function of stripe length at a fixed
photon energy (gain maximum) is fitted in a non-linear regression procedure resulting in the maximum optical gain as fit parameter. Moreover, deviations of the observed curve from the theoretical prediction
indicate the onset of optical gain saturation as important parameter
for optical amplifiers.

III. Results and Discussion

In fig. 2 we depict a measured light output curve with fit. The pump
intensity was well above threshold, leading to an optical gain of a-
bout 370 cm-1. Above a stripe length of 150 μm a deviation from the
exponential law (eq.1) is observed, indicating gain saturation with
stripe length. This result gives the boundary condition for the geome-
tric modelling of optical amplifiers. It is important to point out,
that this result corresponds to the case where the optical signal input is zero and the output is saturated by the ASE itself. In the case
with non-zero input, saturation is reached at even shorter lengths,
thus limiting the size of the OA to this length. Otherwise the effective signal gain depends on the input power of the signal. The saturation of gain caused by ASE only occurs at approximately constant g·L
values for both materials, independent of pump intensity and temperature. The observed g·L values are about 4 in the case of GaInAsP and
2 for GaInAs.
Additionally, the spectral performance of semiconductor lasers depends
strongly on the spectral shape of the optical gain. In fig. 3 we depict ASE spectra and the corresponding gain spectra of GaInAs and
GaInAsP (λ=1.55μm) at 300 K. From the ASE spectra at different excitation lengths we compute the gain spectra as briefly described in section II. Gain saturation was avoided by the appropriate choice of
stripe lengths. From the shape and energetic positions of the ASE
spectra we ascribe the observed emission to band-to-band transitions
for both materials in the whole temperature range (83 K to 300 K) investigated. The full width at half maximum of the gain spectra at
300K for GaInAs is 30 meV and 60 meV for GaInAsP.

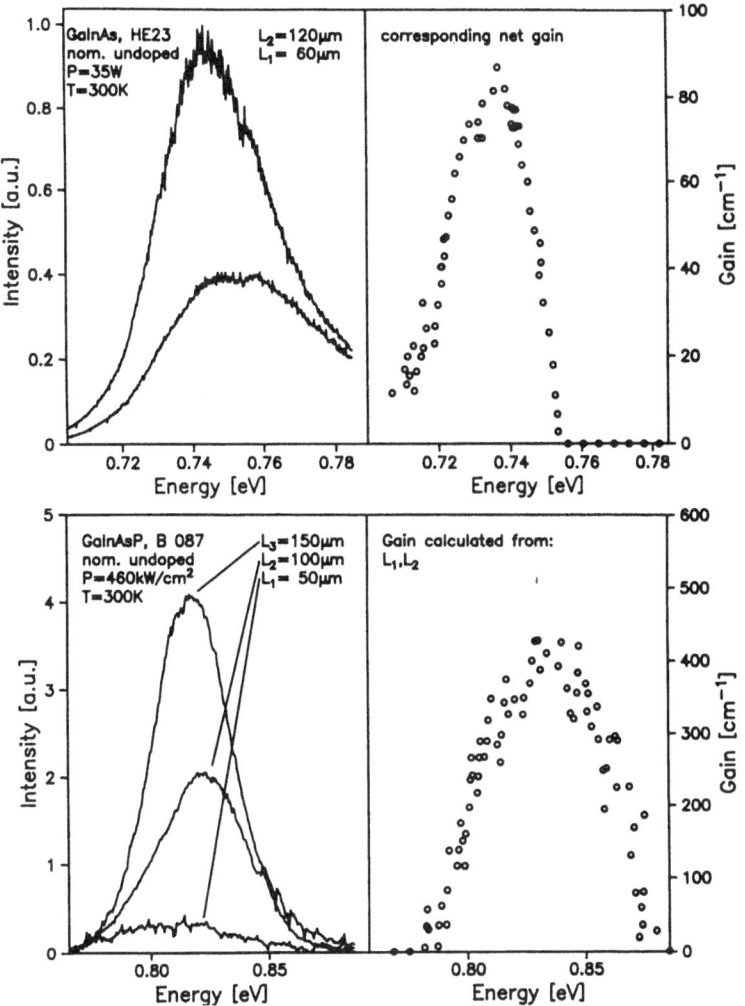

fig.3 ASE at different stripe lengths with corresponding gain
coefficients

To obtain reliable device modelling of OA's and lasers we have to know
the maximum unsaturated optical gain coefficients as function of pump
intensity and temperature. In fig. 4 we show the dependence of the
unsaturated gain on pump intensities at various temperatures.
Above threshold we observe a steep increase of gain with increasing
pump intensity, leading to a maximum gain value which couldn't be enlarged even at highest pump intensities. As shown in fig. 5 the maximum gain values are constant at low temperatures and show a steep decrease above 210 K. At room temperature the observed gain reaches values of 120 cm[-1] for GaInAs and 400 cm[-1] for GaInAsP. These are still
suitable values for optical amplifiers. From our measurements we also
determined the threshold pump intensity for optical gain as a function of temperature. The results are shown in fig. 6 for the case of
GaInAs and GaInAsP. Similar to 1.3 µm quaternary lasers we ob-

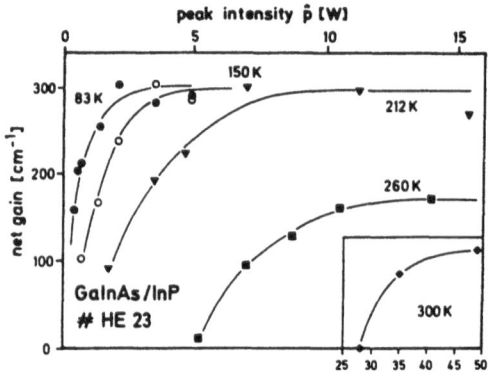

fig.4 net gain as function of pump intensity and temperature

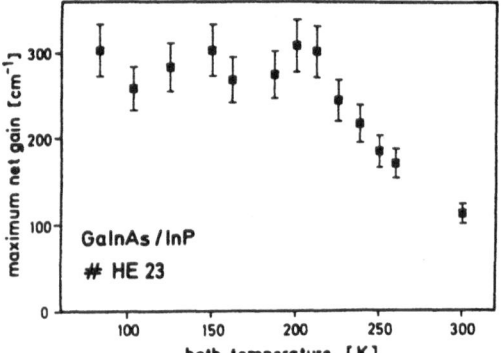

fig.5 maximum gain values as function of temperature

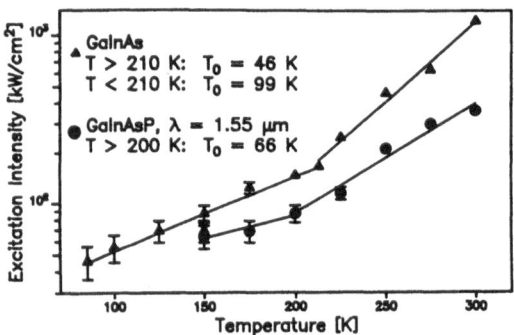

fig. 6 comparison of threshold curves for GaInAsP and GaInAs as function of temperature

serve a breakpoint behaviour in both cases. The T_0 values are 46 K for GaInAs above 210 K and 66 K in the case of GaInAsP above 200 K. The reduced T_0 value at room temperature of the material with the longer wavelength indicates stronger Auger losses and intervalence band absorption (IVBA) as in the case of 1.3 µm GaInAsP lasers /6/. From the breakpoints at equal temperatures in the gain and threshold curve respectively, we find evidence that both are caused by the same physical loss mechanisms, mainly Auger losses and IVBA losses.

IV. Conclusions

We present measurements of optical gain and its saturation in GaInAs and in GaInAsP (1.55μm). We have shown that by means of optical gain measurements the modelling parameters (P_{pump},L,T,material composition) are derivable. The found g·L product for the materials investigated shows that earlier theoretical calculations overestimate the possible optical gain, thus increasing the number of repeaters necessary. Furthermore,temperature-dependent measurements of gain and threshold show evidence that intrinsic material dependent loss mechanisms are responsible for the reduction of the optical gain at higher temperatures for both materials.

V. References

/1/ J.C.Simon, J. Opt. Commun. 4, 51 (1983)
/2/ E.O.Göbel, O.Hildebrand, K.Löhnert, IEEE J. Quantum Electronics QE-13, 848 (1977)
/3/ Y.Yamamoto, IEEE J. Quantum Electronics QE-16, 1073 (1980)
/4/ E.O.Göbel, Photoluminescence and Optical Gain of GaInAsP, in: GaInAsP Alloy Semiconductors, ed. T.P.Pearsall (John Wiley & Sons 1982) pp. 313-338
/5/ Y.Horikoshi, Temperature Dependence of Laser Threshold Current, same vol. as ref. 3, pp. 379-411
/6/ M.H.Pilkuhn, Proceedings of the International Electronic Devices and Materials Conference 1984, Hsinchu, Taiwan, p. 387
/7/ H.Eisele, W.Körber, K.W.Benz, Electronics Letters 19, 1035 (1983)

Monolithic Integration of a Laser and Driving Circuits in GaAs/GaAlAs System for High Speed Optical Transmission

M. Hirao, S. Yamashita, T.P. Tanaka, and H. Nakano

Central Research Laboratory, Hitachi, Ltd., Kokubunji, Tokyo 185, Japan

The monolithic integration of a GaAlAs/GaAs multiquantum well laser with a FET driving circuit on a semi-insulating substrate is described. The circuit operates at more than 2 Gbit/sec.

1. Introduction

It is expected that monolithic integration technology of opto-electronic devices and electronic circuits will play an important role in the field of high bit rates optical communication systems in the furture (1). Several opto-electronic integrated devices have been reported (2-5). For realization of practical opto-electronic integrated circuits (OEICs), a homogenious and large area crystal growth technique is essential. In addition, a new technique for a laser cavity mirror formation without using the conventional cleaving method are also required. The MOCVD or MBE method for crystal growth and the dry etching for mirror formation seem to be promizing for these purposes.

In this paper, the fabrication techniques and the characteristics of a monolithic OEIC device which includes a laser and its driving circuits are described.

2. Structure of Integrated Devices

Laser Structure. Figure 1 shows the cross sectional view of the basic integrated laser structure. A self-aligned structure (SAS) laser with a multi-quantum well (MQW) active layer were formed in a groove on a semi-insulating GaAs substrate by using the two step MOCVD method (6). After the formation of the laser structure, the multi-layers grown on the outside of the grooves were chemically etched off to expose the semi-insulating GaAs surface, where electronic circuits were formed. The reduction of the heights is important for the fabrication of fine gate patterns of high speed field effect transistors (FETs). The partially exposed n^+-GaAs layer in the groove is used as the n-side electrode of the laser, and it is connected to the electronic circuit by wiring metal. The back and the front side facets of the laser cavity were formed by the reactive ion beam etching (RIBE) and conventional cleaving, respectively. The cavity length of the laser is 200 μm.

Driving Circuit Design. Figure 2 shows the circuit diagram of the OEIC device. The circuit consists of 10 FETs and 3 resistors as electronic devices. FET Q_1, Q_2, Q_3, operate as a differential switching circuit for driving the laser diode. For reducing the driving voltage amplitude, an input buffer amplifier constructed by FET Q_4-Q_{10} is also designed. The laser driving current of more than 15 mA at the input pulse amplitude (V_{in}) of 0.8 Vpp can be

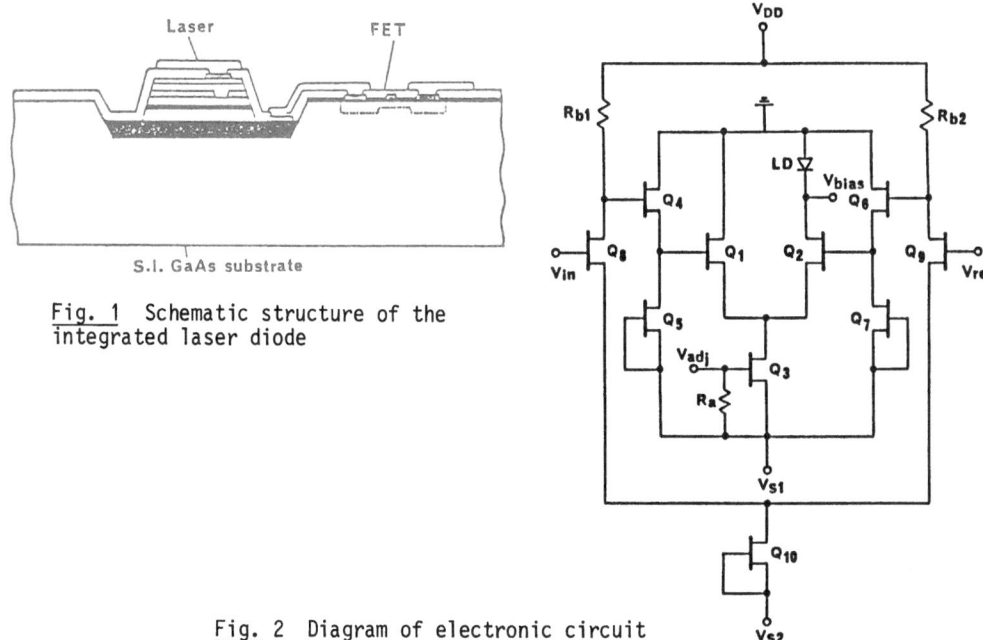

Fig. 1 Schematic structure of the integrated laser diode

Fig. 2 Diagram of electronic circuit

obtained with the circuit design. The driving current can be adjusted by varying the gate voltage of Q_3 (V_{adj}).

Field Effect Transistor. The active layers of FETs were formed by Si ion implantation on the exposed semi-insulating GaAs. The formation of FET electrodes was done by the conventional lift-off technique. Using a high precision photo-lithography, FETs of 1.5 μm gate length were achieved with high reproducibility. The transconductance coefficient as high as 60 mA/Vmm has been obtained at the pinch-off voltage V_{th} = -1.2 V. Figure 3 shows the photograph of a fabricated OEIC chip. The chip size is 1.9 mm × 0.8 mm.

3. Characteristics of Integrated Devices

Laser Diode. Figure 4 shows the I-L characteristics of an integrated laser diode. The threshold current was as low as 40 mA at room temperature. The

Driving Circuit Laser Monitor Output
Detector Circuit

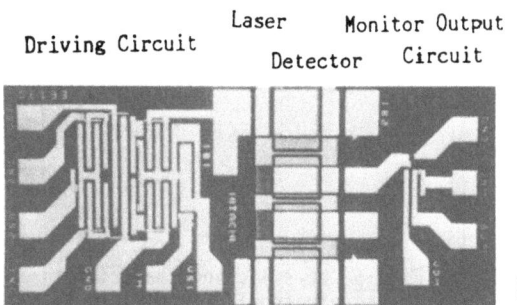

Fig. 3 Photograph of a fabricated OEIC chip

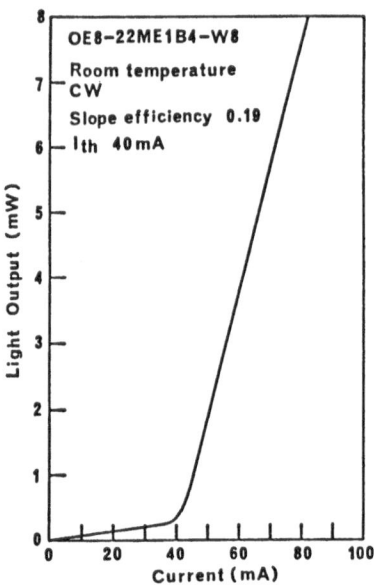

Fig. 4 I-L characteristics of a integrated laser

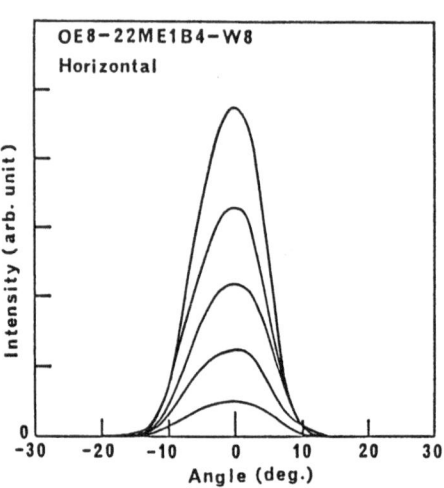

Fig. 5 Far field patterns of a laser parallel to the junction plane

slope efficiency of 0.19 mW/mA was obtained. Compared with the conventional cleaved facets laser with the identical SAS laser structure, the slope efficiency of the laser with a dry etched facet is inferior about 20 %. However, the slope efficiency of 0.19 mW/mA is still enough for practical use, because no difference in threshold currents was observed. Figure 5 shows the far field patterns parallel to the junction plane. No drastic change in the beam angle was observed up to 8 mW. This results indicates that the laser operates at stable transverse mode.

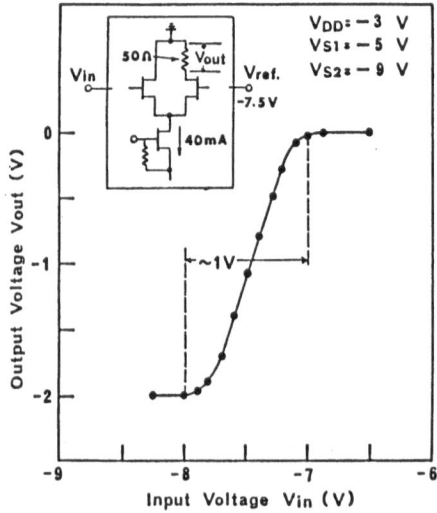

Fig. 6 DC switching characteristics of the electronic circuit

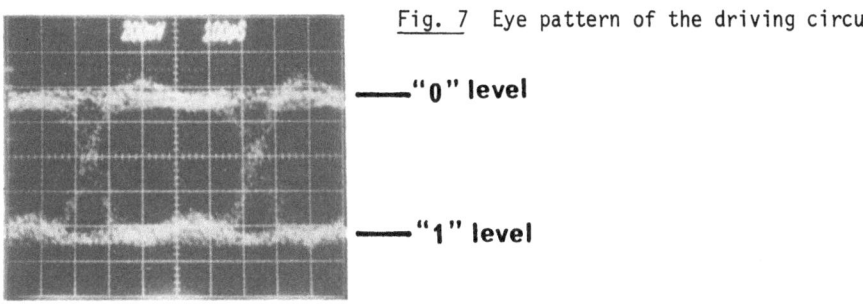

Fig. 7 Eye pattern of the driving circuit

——"0" level

——"1" level

2 Gb/s NRZ

Rise time   130 psec
Fall time   120 psec

Driving Circuits. The electrical response characteristics of the driving circuits were measured by using a chip which included a dummy resistor instead of a laser diode. Figure 6 shows the DC switching characteristics of the circuit. In this experiment, the amplitude of switched current was 40 mA which was determined by the drain saturation current of FET Q_3. The switching voltage was 1.0 V for 40 mA switching and it became about 0.6 V for 15 mA. Figure 7 shows the eye pattern when the driving circuits are operated at 2 Gb/s (NRZ). The rise and fall time of the response waveforms were 130ps and 120ps, respectively. These results indicate the circuit operates up to more than 2 Gb/s. The results of the co-operative operation will be presented.

Acknoledgements. The work was supported by the Agency of Industrial Science and Technology, MITI, Japan, under the National Research and Development Project "Optical measurement and control systems".

References

1. N.Bar-Chaim, I.Ury, A.Yariv: IEEE Spectrum, *19*, 38-45(1982)
2. H.Matsueda, M.Nakamura: "Integrated optoelectronic structure", IEDM Proceedings, Washington, USA (1981)
3. A.Yariv, S.Margalit: "Monolithic optoelectronic integration in semiconductors". IOOC 1983, Technical Digest, 182-183, Tokyo, Japan.
4. H.Matsueda, S.Sasaki, M.Nakamura: "GaAs optoelectronic integrated light sources", IEEE J. Lightwave Tech., LT-1, 261-268 (1983)
5. K.Kasahara, J.Hayashi, H.Nomura: "Gigabit per second operation by monolithically integrated InGaAs/InP LD-FET", Electr. Lett., *20*, 618-619 (1984)
6. K.Uomi, S.Nakatsuka, T.Ohtoshi, Y.Ono, N.Chinone, T.Kajimura: "High-power operation of index-guided visible GaAs/GaAlAs multiquantum well lasers", Appl. Phys. Lett., *45*, 818-820 (1984)

Self-Consistent Analysis of Waveguiding in Phase-Locked Array Lasers

J.P. Van de Capelle, R. Baets, and P.E. Lagasse

Laboratory of Electromagnetism and Acoustics, Sint-Pietersnieuwstraat 41
B-9000 Gent, Belgium

The coupling mechanism between active waveguides is mode-
led in a self-consistent way. The method is applied to a
phase-locked array laser with individual stripe pumping.

1. Introduction

During the last years an increasing amount of effort has been put into the
investigation of phase-locked arrays of semiconductor lasers [1]-[4].
These consist of a number of parallel laser stripes close enough for optical waveguide coupling to occur. Very high power outputs have been obtained, using an array of 40 lasers [2]. A phase-locked array of two electrically independent lasers, showing single longitudinal mode behaviour
has been fabricated [3]. These lasers are therefore promising devices for
high data rate optical communication systems. Moreover, it is possible to
select the fundamental lateral mode, by proper tailoring of the gain distribution accross the array [4]. Such fundamental supermode oscillation
is essential in order to obtain single lobe emission. It requires that
the lasers are individually adressable in order to obtain the proper
gain distribution. More recently, it has been possible to integrate more
lasers by means of a two-level metallisation [13].

To our knowledge, a self-consistent, static model for phase-locked
arrays is presented here for the first time. It allows for the calculation of the operating characteristics at currents ranging from threshold up to multi-lateral mode injection levels.

We have found the threshold fieldpatterns to be highly sensitive
upon the current combination injected in the contacts. Moreover, it will
be shown that for certain current combinations the threshold field has a
planar phase front along the junction plane, which means that the laser
output field does not show astigmation, as is the case for gain-guided
lasers.

Our analysis is based upon the Beam Propagation Method [11], [12].

In the next section we will briefly describe the calculation method.
Next, some examples will be shown and conclusions will be drawn.

2. Calculation method

The analysis of waveguiding in semiconductor lasers is different from
that of passive waveguides in the sense that the refractive index structure is dependent on the optical power density. Moreover, a resonance
condition needs to be fulfilled. The use of the Beam Propagation Method
(BPM), which is well established for passive waveguide structures, is
relatively new in the analysis of lasers [5] - [8]. The main advantage
of the application of the BPM is that it can be adapted quite easily to a
wide variety of structures, and that it allows for both longitudinal and
non-longitudinal calculations.

In all laser modeling methods a number of assumptions are applied, leading to a set of equations which relate different dependent variables.

In our model, an imposed form of the lateral current spreading is used. For a single stripe laser an analytic expression for the injected current density has been derived in [9]. We have added these expressions, which assumes that the distance in between the contacts is large compared with the spreading length of the injected current density. The electron concentration is determined from a simple diffusion equation [6], [8]. The stimulated emission term in this equation is proportional to the average power density, which is calculated as the sum of the lateral mode power densities.

For the TE- field equations we apply the effective index method as described in [10].

In a longitudinal calculation, an adequately chosen starting field is propagated through the dielectric structure. At each step the refractive index is calculated from the optical power density (which is the sum of forward and backward propagating power densities). When arriving at a mirror, the reflected field is calculated and propagation is continued backward. This cycle is continued until convergence is achieved. In a non-longitudinal calculation, the mirror losses are spread over the cavity, which means that the propagation is continued (without reflections) until convergence is again achieved. A problem arises when the field consists of several lateral modes. As these actually lase at slightly different wavelengths, the total power density as used in the diffusion equation consists of the sum of the power densities of the separate modes. However, these separate modes are not available in the calculation. If it is assumed that there are only two modes, and that they are even and odd respectively, a separation of the field into these modes can easily be done [7]. However, this is quite risky since even in a symmetric laser geometry, assymmetric modes may occur well above threshold due to assymmetric index distributions. In a non-longitudinal calculation, the total power density can be found by averaging the "instantaneous" field power density over the propagation distance. Although the propagation needs to be done over a distance which is long compared to the coupling lengths between the modes, this method works well for multilateral laser analysis with two or more modes. The iteration is then finished when the averaged power density converges to a constant profile. In multiple stripe lasers, multimode operation starts at currents only little above threshold. Therefore we have restricted ourselves here to a non-longitudinal calculation.

3. Calculation examples

In a first example we have considered a structure with three stripes, separated by 1.5 μm. The two most outer stripes have a width of 3 μm, the inner stripe has a width of 6 μm.

In fig. 1 we have depicted the threshold current variation of the outer stripes (I_1 and I_3) as a function of the current injected in the inner stripe (I_2) for the even and odd modes of the cavity (curves 1 and 2 respectively). It should be noted that the threshold currents of the even mode are determined in the absence of the odd mode (even if this one is already above threshold) and vice versa. For $I_1 = I_3$ smaller than 43 mA convergence problems occurred for the odd mode. We notice that when the three currents are equal, the even mode reaches threshold for smaller values of these currents. Therefore, this mode will be the first to lase, and it will deplete the electron concentration. As a consequence the threshold current for the odd mode will increase. This is illustrated in fig. 2, which depicts the light current characteristic of this laser when $I_1 = I_2 = I_3$. Around 56 mA we see a typical kink in this curve, indicating the onset of the second order lateral mode. The calculation

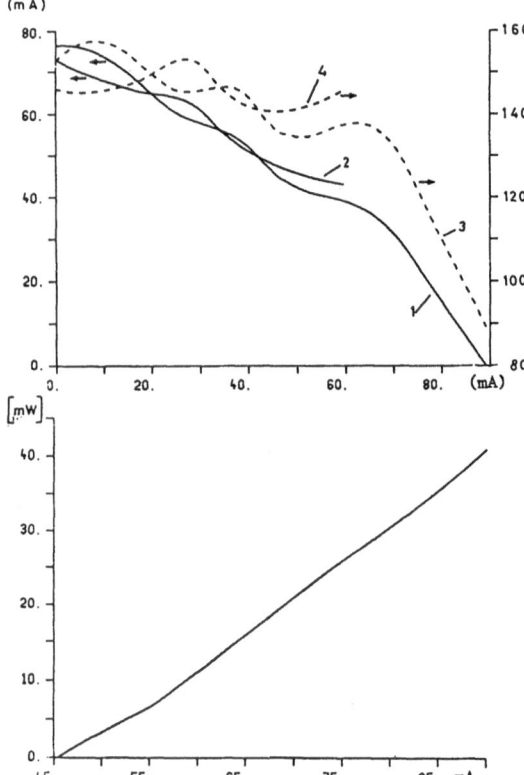

Fig. 1
Threshold current combinations for the even and odd modes ((1),(2)) and the total injected current for both cases ((3),(4)), of the first laser.

Fig. 2
Light-current characteristic of the first laser, when $I_1 = I_2 = I_3$.

therefore required the application of the method described in section two.

At threshold (for $I_1 = I_2 = I_3$) the near field pattern consists of a single lobe with a planar phase front over a large area. These properties remain true over a certain current region and they are, of course, a consequence of the real and imaginary index variations due to the free carrier concentration.

When most of the current is injected in the inner contact, the odd mode has a lower total threshold current than the even mode (see fig.1). In fig. 3 the odd threshold field (amplitude and phase) is drawn for the inner current $I_2 = 0$. We observe some peaks in between the two main lobes. These peaks are the result of the interference pattern of the two main lobes, each of them having a rapidly varying phase, which is correlated to the anti-guiding parameter C. When this parameter is decreased, the interference peaks will diminish. They will also diminish if the stripes are placed closer together.

For the second example we have taken a laser with three stripes, again separated by 1.5 μm, but with a center contact of 3 μm and an outer contact of 1 μm wide. Fig. 4 depicts again the isopower lines P=0 for the even and odd modes (curves 1 and 2 respectively) in the absence of any lasing mode. Curves 3 and 4 give the total injected current at threshold $(I_1+I_2+I_3)$ for both cases. In case of the even mode, we notice there is a relative minimum in the total current $(I_1+I_2+I_3)$ for $I_1=I_2=I_3$ and there is a relative maximum for $I_1 \approx I_3 \approx 23$ mA and $I_2 \approx 76$ mA.

114

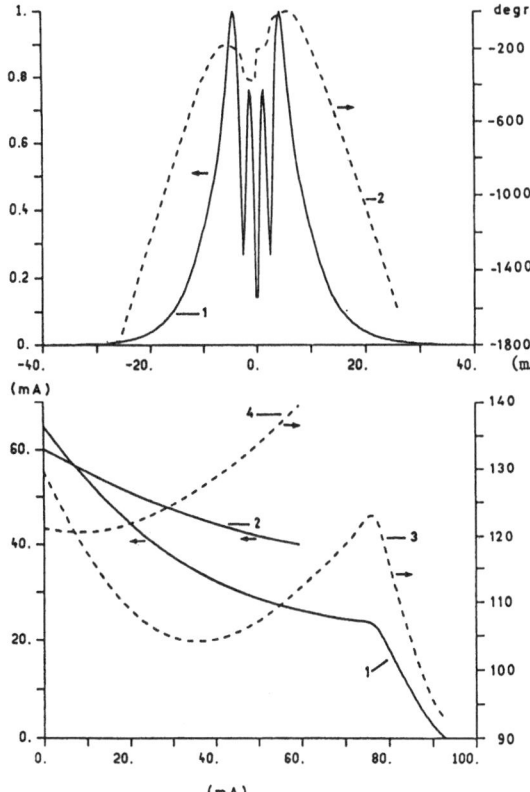

Fig. 3
Threshold
field pattern
(amplitude (1)
and phase (2)),
of the first
laser, when
$I_2 = 0$.

Fig. 4
Threshold cur-
rent combina-
tions for the
even and odd
modes ((1),(2))
and the total
injected cur-
rent for both
cases ((3),(4))
of the second
laser.

Around this current region,the fields become slightly double lobed and
the phase varies rapidly.

4.   Conclusion

We have presented a self-consistent, static, lateral model of individually addressable arrays of semiconductor laser devices operating near
threshold and in multi-lateral mode regime. It has been shown that by
proper tailoring of the lateral gain distribution, we can obtain a
planar phase front along the junction plane, associated with single
lobe emission.

Acknowledgement

J.P. Van de Capelle wishes to thank the Belgian National Fund for
Scientific Research (NFWO) for financial support.

References

[1] J.K. Butler, D.E. Ackley and D. Botez, Appl. Phys. Lett., vol. 44,
      13, pp. 293 - 295, 1984.
[2] D.R. Scifres, R.D. Burnham, C. Lindström, W. Streifer, and T.L.
      Paoli, Appl. Phys. Lett 42 (8), pp. 645-647, 1983.

[3] B.F. Levine, R.A. Logan, W.T. Tsang, C.G. Bethea, F.R. Merrit,
      Appl. Phys. Lett., vol. 42 (4), pp. 339-341, 1983.

[4] E. Kapon, C. Lindsey, J. Katz, S. Margalit and A. Yariv,
      Appl. Phys. Lett. 45 (3), pp. 200-202, 1984.

[5] G.P. Agrawal, W.B. Joyce, R.W. Dixon and M. Lax,
      Appl. Phys. Lett. 43 (1), pp. 11-13, 1983.

[6] R. Baets and P.E. Lagasse, Electron. Lett.,20(1), pp.41-42, 1984

[7] P. Meissner, E. Patzak and D. Yevick, IEEE J. Quantum Electron.,
      vol. QE-20, Nr 8, pp. 899-905, 1984.

[8] R. Baets, J.P. Van de Capelle, P.E. Lagasse, To be published in the
      IEEE J. Quantrum Electron, special issue on semiconductor lasers.

[9] H. Yonezu, I. Sabuma, K.Kobayashi, T. Kamejina, M. Ueno and Y. Nan-
      nichi, Jap. J. Appl. Phys., 12, pp. 1585-1592.

[10] J. Buus, IEEE J. Quantum Electron., 18 (7), pp. 1083-1089, 1982.

[11] J. Van Roey, J. Van der Donk, and P.E. Lagasse, J. Optical Soc. of
      America, vol. 71, Nr 7, pp. 803-810, 1981.

[12] P.E. Lagasse, R. Baets, Nato ASI Series, Serie E : Applied Sciences
      - NR. 86, edited by L.B. Felsen, pp. 375-393, 1983.

[13] J. Katz, E. Kapon, C. Lindsey, S. Margalit, A. Yariv, Electron.
      Lett. 19 (17), pp. 660-662, 1983.

GaAs-GaAlAs Y-Branch Interferometric Modulator

P.M. Rodgers

British Telecom Research Laboratories, Ipswich, IP5 7RE, United Kingdom

Y-branch interferometric modulators have been fabricated with a GaAs-GaAlAs heterostructure, using dry-etching techniques. An extinction ratio of 16.9 dB with 14.6 volt bias was obtained.

1 Introduction

The development of high-speed electro-optic waveguide modulators is of increasing importance for wideband single-mode optical systems operating in the range 1.3 µm to 1.6 µm. External amplitude modulation using a Mach-Zehnder interferometer could be of particular importance because of difficulties in modulating semiconductor lasers directly at high bit rates, due to wavelength chirping, transient oscillations and the increase of the number of longitudinal modes. In addition, this device can form the basis of a frequency shifter[1] and a number of optical logic components[2].

Devices fabricated in III-V semiconductors offer the possibility of integration with lasers and other active components to produce a monolithic optical circuit. Results are presented here of amplitude modulation measurements on a Y-branch Mach-Zehnder interferometer, fabricated in a GaAs-GaAlAs heterostructure.

2 Design and Fabrication

The structure of the modulator is illustrated schematically in figure 1. A GaAs-GaAlAs heterostructure is used to provide vertical optical confinement, and incorporates a $Ga_{0.75}Al_{0.25}As$ buffer layer of thickness ~ 0.5 µm to ensure isolation of the optical wave from the absorbing effect of the metal electrodes. The doping profile results in a good overlap between optical and applied electrical fields and thus a low modulation voltage. Keeping all doping levels less than 5×10^{17} Cm^{-3} minimises loss due to free carrier absorption, and hence propagation loss. The layers were grown by MOCVD at the SERC Central Facility for III-V Semiconductors at Sheffield University. This growth technique provides the large areas of material of good morphology and uniformity that are required for integrated optical devices. Ribs with a width W = 3 µm have been etched to produce single-mode guides. A rib height H = 1.1 µm was used for the straight modulator sections, whilst the height was increased to H = 1.5 µm in the Y-coupler bends to prevent radiation loss. The Y-couplers were formed from two S-bends of radius R = 5000 µm, which produced a tapered input guide of length ~ 120 µm. The heterostructure provides tight lateral confinement, and allows the use of bends with small radius of curvature and low loss, to produce a device of only 7 mm total length.

117

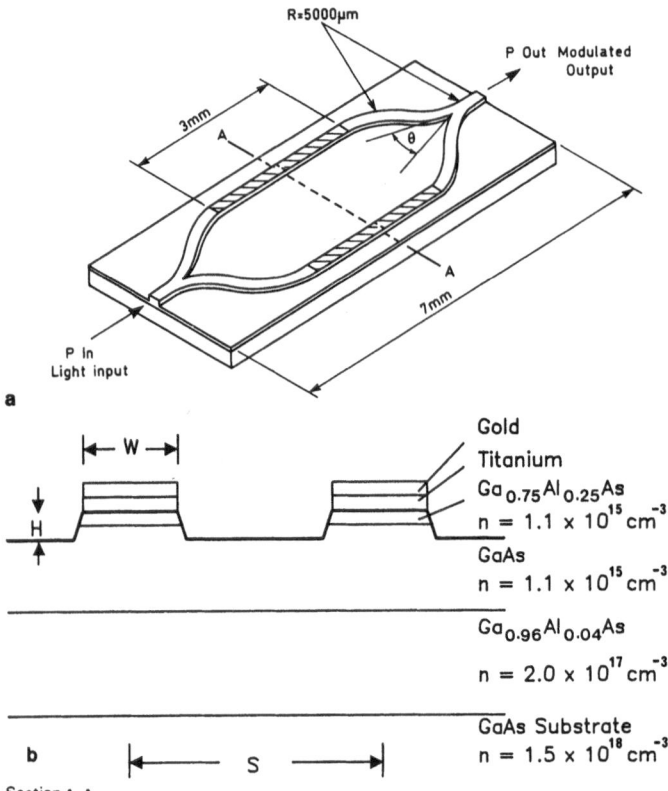

Fig 1a Schematic diagram of a GaAs-GaAlAs Y-branch interferometer
      b Layer structure of a GaAs-GaAlAs rib waveguide

A small branching angle (in practice $\theta < 2°$), and the taper were
designed to avoid mode conversion effects at the Y-junction. The separation
of the arms of the modulator, 20 μm, was sufficient to prevent coupling
between the straight guides in the modulation section.

The fabrication technique is similar to that previously reported[3]. A 3 μm
stripe pattern is defined in Ti-Au using a chlorobenzene/resist lift-off
process, and is used (with a further Ti layer) as a mask for Ar-ion milling.
Rib waveguides are formed with self-aligned electrodes on top. Areas of
metal over the Y-couplers are subsequently removed by a combination of
reactive-ion etching in CF_4 and further Ar-ion milling, which allows the
rib height to be increased in the curved-guide sections. Dry etching
techniques are used throughout, as they have been found to provide a
reproducible and controllable route to fabrication. The process can also
be used to produce high performance directional couplers and phase modulators
by defining the appropriate electrode pattern.

The operation of the device is similar to that previously reported by
Martin[4] for a ZnSe Mach-Zehnder. Light is launched into a single-mode guide
of length 550 μm and split into the two arms by a 3 dB Y-coupler. The
metallised arms form Schottky barriers to the semiconductor, one of which

is reverse biased to apply an electric field in the guiding region. The refractive index change induced by the electro-optic effect will give rise to a phase difference between the two arms. For zero phase difference, the power is coupled to the zero-order mode of the output guide and gives a maximum intensity. For π phase difference, the power is coupled to the first-order mode of the output Y-coupler. As the output guide will not support this mode, the power is radiated into the substrate, resulting in a minimum.

The optical power output P_{OUT} is given by the expression

$$P_{OUT} = \alpha \frac{P_{IN}}{2} \ (1 + \text{Cos} \ \phi)$$

where P_{IN} is the optical input power, and ϕ is the induced phase-shift. α takes account of any loss which is assumed to be equal in both branches. Rediker and Leonberger[5] have shown theoretically that the operation of a Y-branch Mach-Zehnder interferometer is relatively insensitive to small errors in fabrication which lead to power inequality in the arms feeding the output Y-coupler, and to imbalance in this coupler itself.

3 Results and Discussion

The operation of the devices was measured using a chopped beam from a CW Nd-YAG laser at 1.32 μm and a HeNe laser at 1.15 μm. TE polarised light was end-fire coupled into the cleaved input facets via microscope objectives. The output was focussed onto a Ge photodetector, and measured using a lock-in technique. An X-Y recorder was used to plot optical output power against dc modulation voltage. Alternatively, the output could be imaged on an infrared TV camera to check the mode-shape.

The modulator output at zero bias was observed to be single moded at 1.3 μm and 1.15 μm. As a bias voltage was applied, the output decreased as expected, and remained single moded. (Devices with a shorter cleaved (~300 - 400 μm) output guide showed a significant contribution from the higher order mode of the output coupler, as a bias was applied and produced poor extinction ratios). The dc characteristic of light output with applied bias voltage is shown in figure 2 for λ = 1.15 μm. The output approximately follows a cosine function of voltage, although with considerable deviation near zero bias, where the electrical depletion field crosses the optical field. An extinction ratio of 16.9 dB (98%) is achieved with a switching voltage $V\pi$ = 14.6 volts. A phase bias of π (20v) would be required to obtain the best cosine output. All of the devices measured had extinction ratios in the range 14 dB to 17 dB, indicating that this particular design is not sensitive to small fabrication variations. Although this voltage is a little high for many systems applications the electrode lengths are only 3 mm.

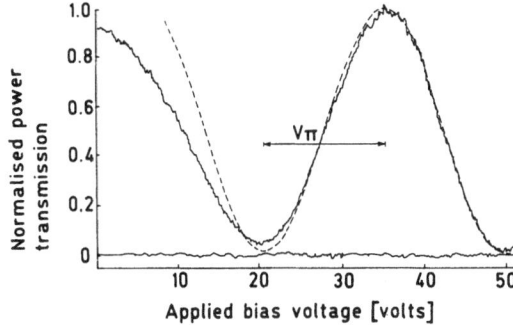

Fig 2 Measured power output with voltage applied to one arm of the interferometer

With the use of longer electrodes, it would be possible to reduce the switching voltage below 10v without increasing the total device length.

The optical loss of these interferometers was compared to that of straight-guide phase modulators of equal length, fabricated on the same chip. The measured output powers were similar to within 1 - 1.5 dB and were limited by input coupling loss. Optical propagation losses of 5.5 dB cm^{-1} have been measured for phase modulators of this type[6]. We conclude that no large attenuation is associated with the Y-coupler or curved-guide sections.

The capacitance per electrode was 1.6 pF, at zero bias. This indicates a potential RC limited bandwidth of a device terminated in a 50Ω system at > 3 GHz. The capacitance could be reduced below 1 pF by eliminating unnecessary bond pads. The use of an electrical bias or reduction of the doping to 10^{14} Cm^{-3} in the guiding region would reduce both the variation of capacitance, and the non-cosine behaviour, with low drive voltage. Optical and electrical transit-time effects should not be apparent at 3 GHz for a device of this length.

4 Conclusion

In conclusion, we have demonstrated guided wave Y-branch interferometers in MOCVD grown heterostructures, using dry-etching to provide a simple and reproducible fabrication procedure. An extinction ratio of 16.9 dB was achieved for ~ 14.6v bias in a device of overall length 7 mm, and which should be capable of modulation to 3 GHz. By careful electrode and semiconductor layer design, it should be possible to reduce the switching voltage to less than 10v, without a speed penalty.

Acknowledgements: We thank J S Roberts for growing the MOCVD slices, M D Learmouth for device processing and A J N Houghton and S Ritchie for useful discussions. Acknowledgement is made to the Director of Research of British Telecom for permission to publish this paper.

References

1 F. Auracher and R. Keil, Appl. Phys. Lett. 36 (8) (1980) pp 626-629.
2 W.D. Bomberger, T Findakly and B. Chen, SPIE Vol 321 Int. Optics 11 (1982) pp 38-46.
3 A.J.N. Houghton and P.M. Rodgers, Second Eur. Conf. on Int. Optics, Florence pp 65-68, (1983).
4 W.E. Martin, Appl. Phys. Lett. vol 26, pp 562-563 (1975).
5 R.H. Rediker and F.J. Leonberger, IEEE J. Quantum Elect. QE18 No 10 (1982).
6 A.J.N. Houghton, P.M. Rodgers and D.A. Andrews, Elec. Lett. 20 (1984) pp 479-481.

High-Efficiency Phase Modulators in InGaAsP/InP

C. Bornholdt, W. Döldissen, D. Franke, J. Krauser, U. Niggebrügge,
H.-P. Nolting, and F. Schmitt

Heinrich-Hertz-Institut für Nachrichtentechnik Berlin GmbH, Einsteinufer 37
D-1000 Berlin 10, Germany

For an electro-optical phase modulator in InGaAsP/InP, optimized by modelling calculations, a phase-shift of 8 deg/Vmm
was achieved.

1. Introduction

In recent years the semiconductor material system InGaAsP on
InP has gained increased importance not only for lasers and detectors for long wavelengths (λ = 1.3 ... 1.6 µm) but also for
additional functional elements like optical waveguides and mo-
dulators/switches needed for future opto-electronic integrated
circuits. The basic component for the switching of light is the
electro-optical phase modulator. For the above mentioned material system we have already presented a rib waveguide phase modulator with a quasi-Schottky contact exhibiting a phase-shift
of 1 deg/Vmm [1]. Similar performance was achieved by FUJIWARA
et al. [2] who employed a double-hetero structure.

Here we report on results of modelling a phase modulator structure with high efficiency and the realization of a corresponding
device with a large phase-shift. The technology applied is well
suited for integration.

2. Design Considerations

The main design criteria of this study were high electro-optical phase-shift per unit length and voltage, low excess loss,
and compatibility with passive rib waveguides suitable for high
intensity transmission [3].

The change of the refractive index Δn_{EO} induced by an electric
field depends on the direction of the field and the light propagation with respect to the crystal axes, the wavelength, and
the band gap of the material. For a sufficiently large separation between the band gap and the photon energy, i.e.
$\Delta E = E_Q - E_{Ph} > 0.2$ eV, only the linear (Pockels) and the quadratic (Kerr) term contribute to the electro-optic effect.
With the electric field perpendicular to the (001) plane we have

$$\Delta n_{EO} = \frac{1}{2} n_Q^3 \left[r_{41} \tilde{E} \pm R_{Kerr} \tilde{E^2} \right]$$

for [110] and [1$\bar{1}$0] propagation of the guided TE mode, respectively [4]. n_Q is the refractive index of the quaternary (wave
guiding) layer, $\tilde{E}$ and $\tilde{E^2}$ are the linear and quadratic electric
field strengths weighted with the intensity distribution of the
guided wave.

The Pockels coefficient was previously measured [5] to be

$$r_{41} = -1.4 \times 10^{-10} \text{ cm/V}$$

and the Kerr coefficient was found to obey the empirical relation [6]

$$R_{Kerr} = -1.5 \times 10^{-15} \exp(-8,85 \ \Delta E \ [eV]) \ cm^2/V^2.$$

It follows that a maximum phase-shift $\Delta \Phi \propto \Delta n_{EO}$ is achievable for a waveguide in [110] direction, where the linear and quadratic terms add up. The choice of the composition of the quaternary layer represents a trade-off between the exploitation of the Kerr effect and the avoidance of electro-absorption (Franz-Keldysh effect), the latter of which was found to be negligible at more than 0.25 eV away from the absorption edge.

In order to obtain the strongest electro-optical effect, it is essential to maximize the overlap integrals for the linear and quadratic contributions ($\tilde{E}$, $\tilde{E}^2$). The field distribution of the guided wave depends on the film thickness t_F and the index profile of the substrate/film/superstrate configuration at the used wavelength λ_{Ph}. The distribution of the externally imposed electric field is determined by the location of the electrode in depth, t_D, and the doping concentration in the quaternary film and in the substrate. Here, the electrode is defined as the equipotential surface (e.g. a Schottky contact or a p^+-n junction) at the upper border of the depletion region. The overlap of the fields is illustrated in fig. 1 for two different optical field distributions, i.e. for a symmetric and an asymmetric waveguide, and for two different electrode locations, namely on top and in the inside of the quaternary layer. Obviously the strongest overlap occurs in the case of an asymmetric waveguide in conjunction with an embedded electrode.

This qualitative argumentation is confirmed by theoretical calculations [6], results of which are given here for an operating wavelength $\lambda_{Ph} = 1.3$ μm and a band gap wavelength of $\lambda_g = 0.97$ μm. Calculated for an asymmetric waveguide on a highly doped substrate fig. 2 shows the dependence of the phase-shift (per unit voltage and length) on the thickness of the quaternary layer for different normalized depths $v = t_D/t_F$ of the electrode. For maximum overlap, the mode pattern can be adjusted by choosing the appropriate slab thickness. Moving the electrode from the sur-

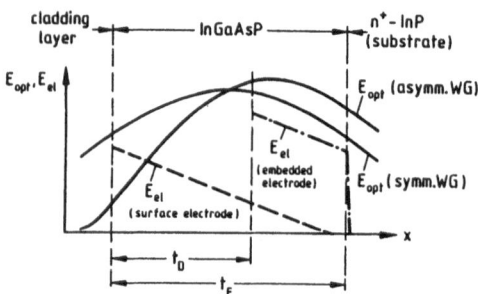

Fig. 1: Overlap of optical and imposed electric field for surface and embedded electrode at the same applied voltage (schematically)

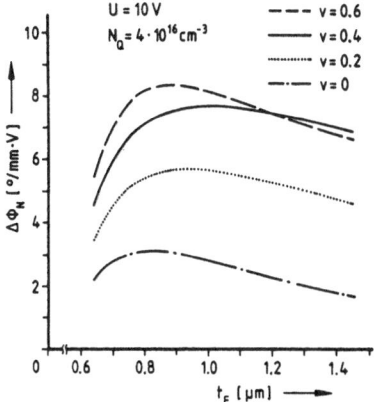

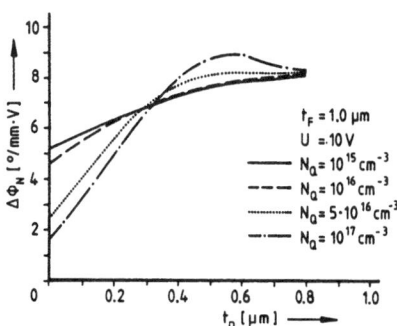

Fig. 2: Calculated phase-shift vs. thickness of quaternary layer for different normalized electrode depths $v = t_D/t_F$ (asymmetric wave-guide)

Fig. 3: Calculated phase-shift vs. electrode depth for different doping concentration of the InGaAsP layer (asymmetric waveguide)

face to approximately the center of the waveguiding film suc-
ceeds in a considerable increase of the phase-shift due to a
"compression" of the imposed electric field into the region of
maximum light intensity. The higher the film doping level the
more pronounced this effect is, as shown in fig. 3, because the
depletion region is narrowed yielding a higher maximum field
strength. Thus higher background doping levels in the film become acceptable where, of course, the maximum voltage will be
limited by the onset of avalanche breakdown.

The capacitance of the structure increases with the narrowing
of the depletion layer; this is, however, counteracted by the
gain in phase-shift per unit device length.

3. Device Technology and Measurements

N-type InGaAsP layers with a bandgap wavelength of 0.97 μm,
background doping concentration of 4×10^{16} cm^{-3}, and thickness
of about 1 μm were grown on (001)-orientated Sn-doped InP substrates by LPE. Waveguide patterns consisting of 4 μm wide
stripes were generated along the [110] direction using a trilevel resist system. This resist structure was utilized first as
an etching mask for delineating rib waveguides (RWG) accomplished by removing approx. 100 nm of the quaternary layer in a
$H_2SO_4/H_2O_2/H_2O$ solution. The trilevel structure secondly served
as a lift-off mask for patterning a thermally evaporated multilayer of 50 nm SiO and 200 nm Al$_2$O$_3$, providing diffusion and
contact windows self-aligned to the RWGs. SiO was found to
be a favourable diffusion mask, but tends to crack when applied
in thickness of more than about 100 nm; therefore Al$_2$O$_3$, which
exhibits satisfactory mechanical stability, was used for rein-
forced field passivation.

A shallow p-n junction (about 0.5 μm in depth) was formed in
the quaternary layer by in-diffusion of Cd through the previously defined windows. The p$^+$-regions were contacted by a CdO

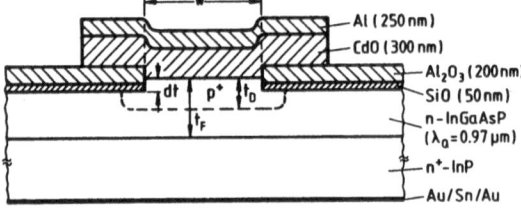

Fig. 4: Cross-section of phase modulator
device

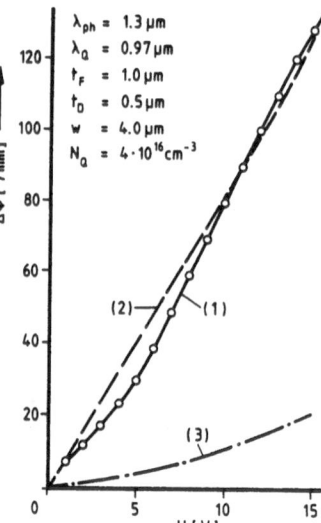

Fig. 5: Measured (1) and calculated (2)
voltage induced phase-shift; curve (3)
gives the calculated quadratic contribution

layer (300 nm). CdO is known as a semi-transparent conductor;
from results of preliminary experiments the parameters of deposition (sputtering with 80% Ar + 20% O_2) and annealing (300°C,
N_2) were adjusted to achieve a reasonable trade-off between
additional optical loss ($\Delta\alpha \approx 1$ cm^{-1} for TM-polarized light at
λ_{Ph} = 1.3 µm) and electrical resistivity ($\varrho \approx 5 \times 10^{-3}$ Ωcm). For
further reducing the series resistance an Al layer (250 nm) was
evaporated onto the CdO; both layers were patterned by lift-off
to define contacting stripes on top of the RWGs. A cross-section
of the completed phase modulator device is given in fig. 4, where
relevant parameters are indicated.

Experimental results are given in fig. 5, where the phase-shift
per unit length $\Delta\Phi$ is plotted versus the applied voltage. For
large signal operation (0 V to 10 V) a phase-shift of 8 deg/V mm
was measured for a typical device. To our knowledge this is the
highest value for RWG phase modulators in the InGaAsP/InP material
system. For comparison the calculated electro-optically induced
phase-shift and its quadratic contribution, which amounts to
more than 10 %, is included in fig. 5. The observed discrepan-
cies at low voltages are not yet fully understood; they may be
partly ascribed to effects of fringing fields and a more complex doping profile than assumed in this model.

Conclusions

The efficiency of an electro-optical RWG phase modulator in
InGaAsP/InP was considerably improved on the basis of a model
aiming at maximum field overlap. A phase-shift of 8 deg/V mm
was measured with a device employing a p-n junction located
near the center of the wave guiding layer.

Acknowledgement

The skilful assistance of A. Döhler, M. Klug, M. Mahnkopf and
I. Tiedke in device fabrication are gratefully acknowledged.
We thank M. Schlichting for computations and P. Albrecht for
optical measurements.

This work was supported by the Federal Ministry of Research
and Technology (BMFT) and the Senate of Berlin (West).

References

1   U. Niggebrügge, W. Döldissen, N. Grote, M. Klug, I. Tiedke,
    M. Schlak, H.-P. Nolting, J. Krauser: NTG-Tagung "Integrated
    Optics", Berlin, Dec. 1983

2   M. Fujiwara, A. Ajisawa, Y. Sugimoto, Y. Ohta: Electronics
    Letters, Vol. 20, No. 19, p. 790, Sept. 1984

3   P. Albrecht, H.G. Bach, C. Bornholdt, W. Döldissen, D. Franke,
    N. Grote, J. Krauser, U. Niggebrügge, H.-P. Nolting,
    M. Schlak, I. Tiedke, R.A. Logan, F.K. Reinhart: 2nd ECIO,
    Florence, Oct. 1983

4   H.G. Bach, J. Krauser, H.-P. Nolting, R.A. Logan, F.K. Rein-
    hart: Appl. Phys. Lett., Vol 42 (8), p. 692 (1983)

5   P. Albrecht, C. Bornholdt, J. Krauser, H.-P. Nolting,
    B. Sartorius: NTG-Tagung "Integrated Optics", Berlin, Dec.
    1983

6   to be published

Optical Waveguide Modulators and Switches in GaAs/GaAlAs Heterostructures

R.G. Walker and M.W. Jones

Plessey Research (Caswell) Limited,
Towcester, Northants, NN12 8EQ, United Kingdom

Optimised electro-optic devices made by MOCVD of GaAlAs/GaAs
have achieved high efficiencies. Mach-Zehnder modulators with
6 mm electrodes have achieved 95 percent modulation for a
5 Volt swing.

1 Introduction

Heterostructures in III-V semiconductors - especially GaAs/GaAlAs - are
increasingly being used to make waveguide based electro-optic devices such
as modulators and directional coupler switches [1,2,3]. Epitaxial growth
processes have improved and become more versatile since the advent of
molecular beam epitaxy (MBE) and metallo-organic chemical vapour deposition
(MOCVD), stimulating research into new applications. The MOCVD process
used in this work is capable of producing the large areas of excellent
morphology required for long optical waveguide devices.

Slab waveguides, consisting of a GaAs/GaAlAs heterostructure,can provide
very strong optical confinement, permitting very efficient electro-optic
interaction with the depletion field of a reverse biased rectifying
junction. Moreover, the aluminium fraction in the cladding layers is
continuously variable and can be graded to tailor the guided mode profile.
Lateral confinement is induced by etching ribs into the upper cladding
layer to produce guides of the strip-loaded type. Figure 1 (below)
illustrates the structure used in this work.

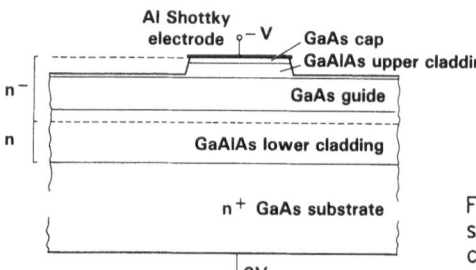

Fig.1. Strip-loaded, double heterostructure, electro-optic waveguide
cross-section

The general potential advantages of III-V semiconductors for
electro-optic devices are well known: monolithic integration with lasers
detectors and fast electronics, and large travelling wave device bandwidth.
However the low (compared with $LiNbO_3$) electro-optic coefficient requires
that the structure be optimised.

We have previously reported modulators and switches of this type [3], however those discussed herein represent a considerable improvement due to the following optimisation procedure.

2 Optimisation

The aim of the optimisation was to minimise the voltage required to achieve a given phase-shift in a phase modulator. To this end, a computer program was written to calculate the optical-power/electric-field overlap (and hence the phase-shift) as a function of voltage. A one-dimensional approximation was used, involving the field distribution of a slab waveguide (up to 4 layers with metallised surface and TE polarisation) and the electric field of a reverse biased, one sided p-n junction. Results from this program have been found to match experimental phase-shift/voltage curves (obtained from mach-zehnder interferometers) very well [4].

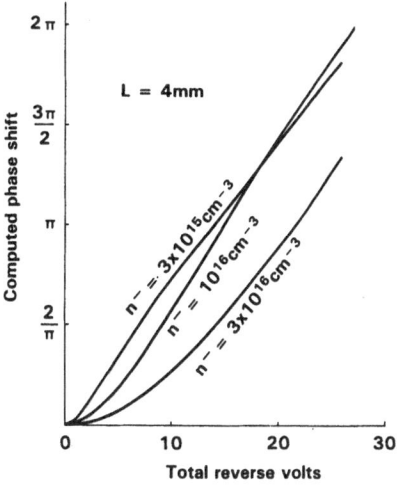

Fig.2.
Computed phase/voltage curves for a (non-optimised) 1.5 μm thick GaAs rib guide, showing the effect of background doping

The phase/voltage curves are found to have the following features (see Fig. 2):

(i)   A non-linear low-voltage region, with voltage offset, due to the small initial depletion depth (less than that of the optical peak).

(ii) A high slope region due to the rapid increase of overlap as the depletion region crosses the optical peak.

(iii) A linear region where the guide is fully depleted through to the n doped ($5x10^{17}$ cm^{-3}) depletion limiter layer. The slope of this region corresponds to that expected for a perfect dielectric with similar overlap, and is usually less than that of region (ii).

The above features are clearly seen for background carrier concentrations of $n^- > 3x10^{15}$ cm^{-3}; however, the curve is almost perfectly linear for $n^- < 5x10^{14}$ cm^{-3}, since the built-in potential of the Schottky barrier is sufficient to fully deplete the guide. As the background doping increases, regions (i) and (ii) become more expansive.

Different optimisation techniques are appropriate for highly doped (n⁻ around 10^{16} cm⁻³) and low doped (n⁻ <10^{15} cm⁻³) cases. In the former, where the high slope region (ii) is the most useful, the above computer program is used to find a specification where the slope is maximised and the offset minimised. In the low doped case the problem is simpler, it being necessary only to minimise the mode depth so that the depletion limiter layer may be placed as near the surface as is consistant with low-loss.

In GaAs/GaAlAs heterostructures there are too many variable parameters to find a unique optimum structure. Refractive indices for the upper and lower cladding were, therefore, chosen and an optimum guide layer thickness obtained for each. The methods for low and high doping were found to give similar optimum values. The upper cladding index was fixed at 3.3 (24% Al by extrapolation from the results of [5]). The various optimum solutions were then compared,and that using 16% Al in the lower cladding layer chosen as being most likely to offer good performance with low loss (i.e. the guide layer, at 0.7 μm, not too thin). The maximum rib etch- depth for monomode guiding was obtained using the effective index method.

3. Device Fabrication

Directional couplers with 4 mm interaction length and mach-zehnder interferometer modulators with 2° (total angle) y-branches and 6mm electrodes were made. The directional coupler waveguides diverge (2° total angle) to 30 μm separation at both ends and guide widths were about 3.5μm with separations of 2, 2.5 and 3 μm. The masks were made by electron-beam lithography.

The waveguide patterns were replicated on the semiconductor surface by lift-off of aluminium about 0.4 μm thick. This then doubled as the etch-mask and (after removal of unwanted regions) Schottky electrodes. Wet chemical etching was used.

4. Mach Zehnder Interferometers

Figure 3 shows a mach zehnder response. The steep slopes at zero volts and the general uniformity of the sinusoid indicate that the background carrier concentration was low (<10^{15} cm⁻³). The V_π value (peak-trough voltage) of 6-7 volts (it increases slightly with voltage) is as expected from theory,assuming an electro-optic coefficient of 1.5×10^{-6} μmV⁻¹.

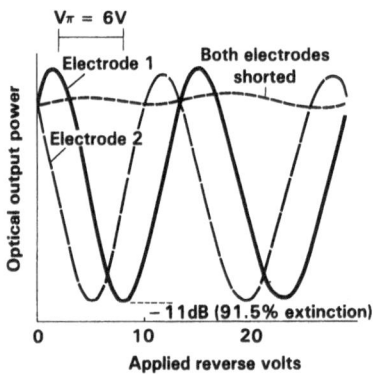

Fig.3.
Recorded response of mach-zehnder interferometer (optimised structure): 6 mm electrodes $\lambda = 1.3$ μm

The difference observed in the response to alternative electrodes indicates a small built-in path difference between the branches. Extinction ratios exceeding 95% have been observed in some devices, though there is at present insufficient control over y-junction quality to reliably give more than 90%. When both electrodes are shorted, there is a residual throughput fluctuation which, in some devices, was very large (e.g. Fig.5). This is attributed to light from the y-junction "point" being confined between the guides and acting as an effective third branch.

5. Directional Couplers

The directional couplers were etched sufficiently that those with 2 μm guide separation had length/coupling-length ratio $L/L_0 \simeq 2$ and those with 3 μm separation had $L/L_0 \simeq 1$. Since L_0 is very sensitive to dimensions there were individual differences between the 12 of each type on the chip. Where $L/L_0 = 1$ uniform $\Delta\beta$ connection is sufficient and reverse $\Delta\beta$ achieves nothing; where $L/L_0 = 2$ the opposite is true. Figure 4 shows the response of a device where $L/L_0 = 2$ (exactly) with reverse $\Delta\beta$ connection. Both switch states were achieved with -20 dB (or better) crosstalk and with equal voltages on the electrode segments. Slightly unbalancing the voltages to compensate device imbalances should improve crosstalk further, though clearly it will greatly simplify use of these devices if such adjustments are unnecessary.

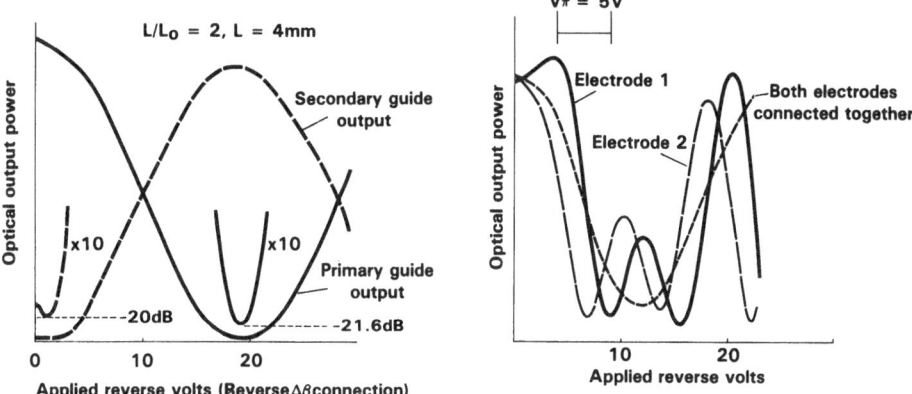

Fig.4. Recorded response of directional coupler switch (optimised structure)

Fig.5. As Fig.3 but using material with higher background doping ($n^- \simeq 10^{16}$ cm^{-1}). The non-ideal response is due to other factors, e.g. y-branch quality

6. Losses

Optical attenuation of these particular devices has not been measured since non-destructive methods are unavailable. However, similar straight waveguides have demonstrated losses of less than 2 dB/cm over short lengths ($\sim$ 5mm).

7. Conclusions

Optical modulators and switches have been demonstrated which approach the maximum efficiency achievable with the GaAs/GaAlAs material system. The efficiency, in terms of phase-shift per volt per unit length, $(5°V^{-1}mm^{-1}$ is demonstrated in Fig.3) can be improved further by using a higher background carrier concentration, thus expanding and making available the non-linear high slope region of the phase/voltage curve (see Fig.2). Figure 5 gives the response of a mach zehnder interferometer made in such material ($n^- \simeq 10^{16}$ cm^{-1}) which was otherwise the same as that of Fig.3. This clearly shows the expected non-linearity at low voltages and steeper slopes at intermediate voltages, producing a V_π value of 5 volts or less and an efficiency of $6°V^{-1}mm^{-1}$. Such highly efficient devices look promising for high speed modulation where large voltage swing is problematical.

Optical switching of 1.3 μm light (with 20 dB crosstalk) over 4 mm length using only 18 volts has been demonstrated.

Future work will concentrate on reducing losses and investigating high speed modulation.

8. References

1. Carenco A. et al: Seventh topical meeting on "Integrated and Guided Wave Optics", 1984 (IEEE/OSA) Proc. No.ThB$_4$-1
2. Houghton A.J.N. et al: Elect. Lett. 20, 479, 1984
3. Walker R.G, Carter A.C: IEEE Workshop on Int. Opt. and Related Tech., 1984 Proc., pp. 103-105
4. Walker R.G, Jones M.W: IEE Colloquium "Heterostructure Opto-electronic and High Speed Devices", 1984, Digest No.1984/93, pp.5/1-5/3.
5. Casey H.C. et al: Appl. Phys. Lett. 24, 63-65, 1974.

Anti-Reflecting Mass Transported Windows for InGaAsP/InP Integrated Lasers

Bjorn Broberg

Institute of Microwave Technology, Box 70033, S-10044 Stockholm, Sweden

Sekartedjo Koentjoro, Kazuhito Furuya, and Yasuharu Suematsu

Tokyo Institute of Technology, Dept. of Physical Electronics, 2-12-1 0-okayama Meguro-ku, Tokyo 152, Japan

Phase-adjusted DFB laser with novel 25μm long mass-transported windows is proposed. Low threshold, single mode operation with asymmetric output efficiency depending on location of the phase adjusting groove is obtained.

Introduction

A future breakthrough of semiconductor integrated optoelectronics is highly dependent on new improved and simplified processing technologies. Mass transport is such a novel process which was first established as a way to make high-performance Fabry-Perot type lasers [1-4]. Recently, this process has also been employed for mirrorless, integrated lasers [5-7]. For such devices it is essential to reduce the reflections from the cleaved facets. The methods reported to date to achieve this, for instance anti-reflection coatings [8] or regrown window structures [9], require additional process steps for mass transported devices.

Therefore, a simple but efficient way to reduce facet reflections is needed to further develop mass transported integrated lasers. We propose mass transported windows to fulfill this need.

Design and related laser fabrication

By a special etch mask design antireflecting windows can be obtained without additional process steps in the fabrication of mass transported lasers. The method is demonstrated for phase-adjusted DFB lasers (Fig.1) where the phase of the optical wave has been adjusted by $\pi/2$ in a groove in order to maximize the single-mode selectivity [6,7]. With waveguide dimensions typical for narrow-stripe DFB lasers, a few μm window length is sufficient to obtain 5% reflectivity while 25μm windows are required to suppress the reflections to less than 1%, assuming diffraction of a Gaussian wave [9].

Such long windows can be made by simply varying the laser stripe width as illustrated in Fig.2a. The stripe width is decreased by 2μm in the 50μm long window regions that separate the 300μm long active regions in the lasing direction. After undercutting etching of the LPE-grown structure a 1-1.5μm wide active stripe remains in the active regions, while the window regions are completely undercut (Fig.2b-d). To monitor the undercutting, 5·50μm sections, i.e. dummy window regions, were made in some

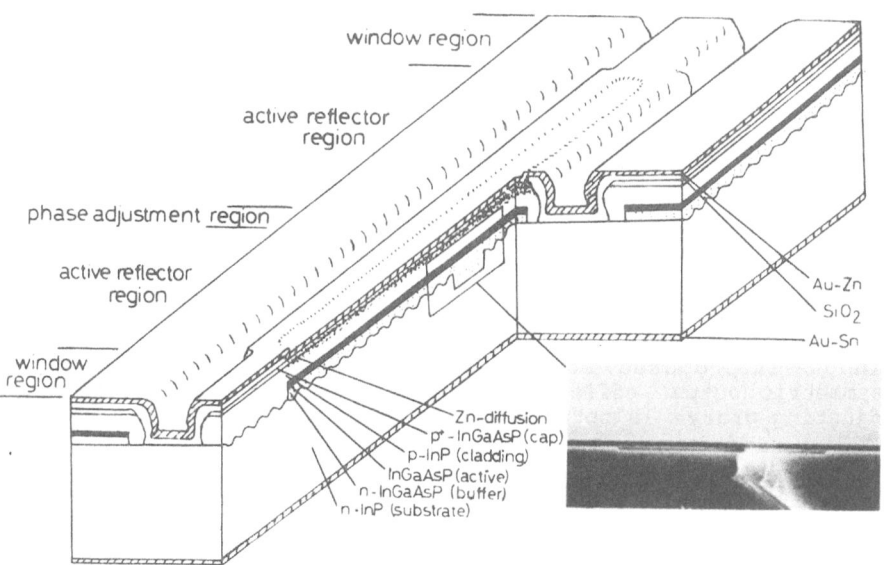

Au-Zn
SiO₂
Au-Sn

Zn-diffusion
p⁺-InGaAsP(cap)
p-InP (cladding)
InGaAsP (active)
n-InGaAsP (buffer)
n-InP (substrate)

Fig. 1 Cut-away view of the laser. The photograph shows the phase-adjusting groove.

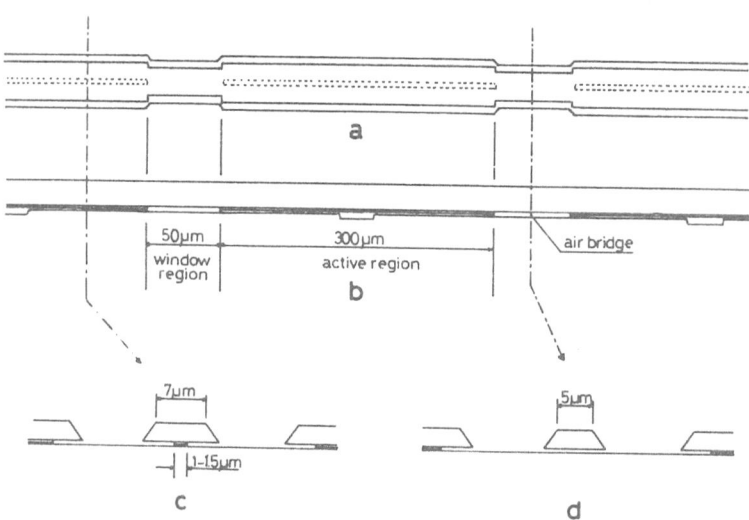

50µm | 300µm | air bridge
window region | active region
b

7µm | 5µm
1-15µm
c | d

Fig. 2 The laser structure before mass transport: (a) top view, (b) side view of the laser mesa, (c) cross-section active region, (d) cross-section window region.

places in the surrounding areas. The etching was interrupted when these sections fell off, indicating that the window regions were fully undercut while desired stripe width remained in the active regions.

The InP air bridges, that are thus formed in the window regions, act as guides for the subsequent mass transport process. After

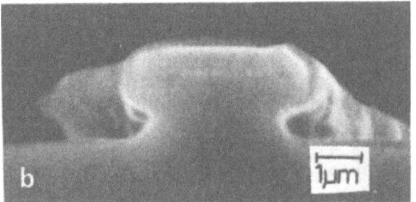

Fig. 3 (a) Cross-section of active region after mass transport, (b) cross-section of the window region 25 µm from the active region. The sample is stain-etched in the same way as in (a).

mass transport under an InP cover and a graphite lid in a conventional LPE system at $700^{O}C$ for 2h in H_2 atmosphere, InP has moved not only to bury the heterostructure (Fig.3a) but also to completely fill the volume under the air bridges (Fig.3b). By cleaving in the middle of the window regions, lasers with 25µm homogeneous InP between the waveguide ends and the cleaved facets are formed.

Laser results

Lowest lasing threshold current was 38mA. Fabry-Perot mode oscillation was efficiently suppressed by the windows. Single mode lasing was obtained, while subthreshold spectra show uniform and symmetric mode distribution, indicating that mode degeneracy is avoided so that the single mode selectivity is maximized.

Lasers with the phase adjusting groove displaced from the cavity center show higher output efficiency from the short than from the long reflector. For extreme displacements the mode selectivity is deteriorated. These observations are in agreement with coupled-mode theory and verify therefore both the effect of the phase adjustment and the efficiency of the mass transported windows.

Conclusions

Mass transported windows is an attractive way to reduce the reflections from the facets in integrated InGaAsP/InP lasers. Without any additional processing to make the windows, the reflections from each facet were reduced to less than 1%. Mass transport was achieved over a length of 25µm, which is about 10 times the previously reported values. Due to the efficient reduction of facet reflections, single-mode selectivity and asymmetric output efficiency of phase-adjusted DFB lasers could be observed in agreement with theory.

Acknowledgement

Mr. F.Koyama, Mr. Y.Tohmori and Mr. N.Eda are gratefully acknowledged for experimental assistance and discussions. This work was supported by a Scientific Research Grant-In-Aid from the Ministry of Education, Science and Culture, Japan, by the National Swedish Board for Technical Development, Sweden and partly by KDD and NTT.

References

[1]   Z.L.Liau, J.N.Walpole: ''A novel technique for GaInAsP/InP buried heterostructure laser fabrication'', Appl.Phys.Lett., Vol.40, No.7, p.568-570, (1 Apr. 1982)

[2]   T.R.Chen, L.C.Chiu, K.L.Yu, U.Koren, A.Hasson, S.Margalit, Y.Yariv: ''Low threshold InGaAsP terrace mass transport laser on semi-insulating substrate'', Appl.Phys.Lett. Vol.41, No.12, p.1115-1117, (15 Dec.1982)

[3]   T.R.Chen, L.C.Chiu, A.Hasson, K.L.Yu, U.Koren, S.Margalit, A.Yariv: ''Study and application of the mass transport phenomenon in InP'', J.Appl.Phys., Vol.54, No.5, p.2407-2412, (May 1983)

[4]   Z.L.Liau, J.N.Walpole, D.Z.Tsang: ''Fabrication Characterization, and Analysis of Mass-Transported GaInAsP/InP Buried-Heterostructure lasers'', IEEE J.Quant.Electr., Vol.QE-20, No.8, p.855-865, (Aug.1984)

[5]   B.Broberg, F.Koyama, Y.Tohmori, Y.Suematsu: ''1.53µm DFB lasers by mass transport'', Electr.Lett., Vol.20, No.17, p.692-694, (16 Aug. 1984)

[6]   B.Broberg, K.Sekartedjo, F.Koyama, Y.Tohmori, Y.Suematsu: ''Mass transported 1.53µm DFB lasers with improved longitudinal mode control'', 10th ECOC conf. Stuttgart, 3-6 Sept.1984, paper PD-2 (VDE-Verlag,Berlin)

[7]   K.Sekartedjo, B.Broberg, F.Koyama, K.Furuya, Y.Suematsu: ''Active Distributed Reflector Lasers Phase Adjusted by Groove Region'', Jap.J.Appl.Phys., Vol.23, No.10, p.L791-L794, (Oct.1984)

[8]   Y.Itaya, K.Wakita, G.Motosugi, T.Ikegami: ''Phase Control by Coating in 1.5µm Distributed Feedback Lasers'', to appear in IEEE J.Quant.Electr., June 1985

[9]   K.Utaka, S.Akiba, K.Sakai, Y.Matsushima: Effect of Mirror Facets on Lasing Characteristics of Distributed Feedback InGaAsP/InP Laser diodes at 1.5µm Range'', IEEE J.Quant.Electr., Vol.QE-20, No.3, p.236-245, (Mar. 1984)

Totally Reflecting Mirrors: Fabrication and Application in GaAs Rib Waveguide Devices

P. Buchmann, H. Kaufmann, H. Melchior, and G. Guekos
Swiss Federal Institute of Technology, CH-8093 Zürich, Switzerland

Reactive ion etched rib waveguide mirrors in GaAs have low loss and are orientation independent. These properties makes them promising components of I.O devices, especially for improved waveguide to fiber coupling.

1 INTRODUCTION

In integrated optical circuits for communication and optical signal processing applications there exists a need for waveguides with large angle directional changes. Such a bending structure would facilitate single-mode fiber to waveguide coupling and help to reduce the large length-to-width ratio of waveguide chips consisting of several coupled devices.

Rib waveguide bends [1] in n/n+ III-V semiconductor material require large radii of curvature (10 to 30 mm) in order to keep radiation losses low. Abrupt directional changes (corner bends) allow only for small displacement angles and are very space consuming. Totally reflecting waveguide mirrors are an alternative structure. They have been demonstrated on photoelastic waveguides using wet chemical etching [2]. With reactive ion etching (RIE) [3] low-loss rib waveguides [4] and good quality laser mirrors [5] have been fabricated in GaAs/AlGaAs.

We have applied this technique to the fabrication of totally reflecting mirrors in single-mode n/n+ GaAs rib waveguides. The principle of a waveguide with a 90⁰ directional change is shown in Fig.1. An important application of this device is the coupling of single-mode fibers to the parallel waveguides of a directional coupler or to the arms of a Y-junction.

In contrast to anisotropic wet chemical etching, the RIE fabrication method is applicable to any deflection angle, as long as the total reflection condition is maintained.

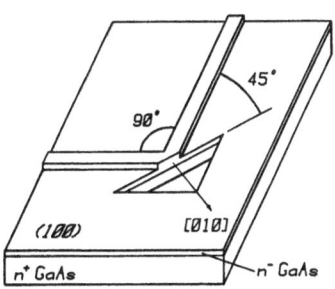

Fig.1 Schematic view of n/n+ GaAs rib waveguide mirror

2 THEORY

In order to evaluate the technological requirements,the influence of a mirror displacement Δ along [010] and of a small mirror rotation angle θ around the optimum 45° position on the losses was calculated. The light was regarded as collimated in the mirror region because the beam divergence is low (1 to 3°). The coupling efficiency between input and output waveguide is defined as the overlap integral of the modes in the mirror plane. The field distribution E(x,y) of a 6.4 μm wide rib waveguide was calculated using a finite difference method. Performing the necessary coordinate transformations,we find for the coupling efficiency [6]:

$$\eta = (1+2\theta) \ \frac{\left| \int\int E(x,y)E(x,(1+2\theta)y - \sqrt{2}(1+\theta)\Delta)\exp[-i\beta(2\theta y - \sqrt{2}\theta\Delta)]dxdy \right|^2}{(\int\int E^2(x,y)dxdy)^2}$$

The results obtained by numerical integration are shown in Fig. 2. Only small angles of rotation and small translations of the mirror are allowed for efficient coupling from input to output guide. The 1 dB power loss limits are 0.6° and 0.9 μm respectively. Therefore, highly precise mask aligning is required in the fabrication process. The limit for a 1 dB power loss caused by a non-vertical mirror plane was estimated to about 2° for the inclination angle.

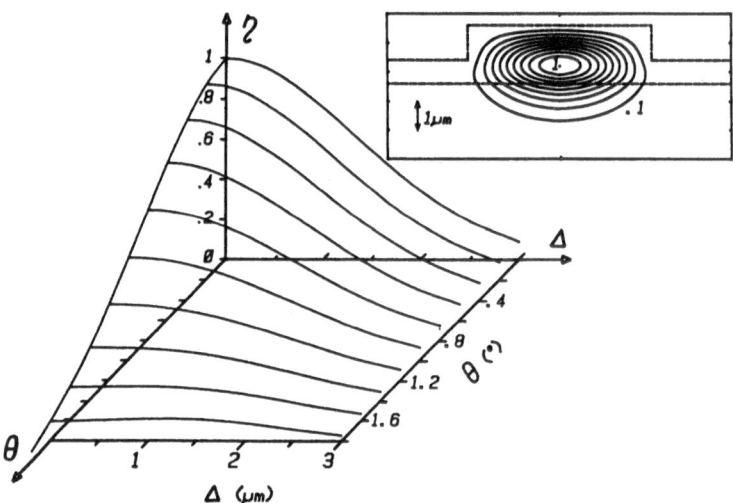

Fig. 2 Calculated intensity distribution and mirror coupling efficiency η for a 6.4 μm wide n/n+ GaAs rib waveguide (Δ translation of mirror along [010], θ small rotation angle)

The phase mismatch between input and output mode has the strongest effect on the coupling efficiency. The phase factor is linearly depending on the lateral coordinate, the coupling efficiency is therefore related to the width of the field distribution. Thus, the mirror loss will be smaller for heterostructure waveguides with their less extended profile as compared to n/n+ waveguides for a given rotation and inclination of the mirror.

3 FABRICATION

The devices were fabricated in VPE- and LPE-grown (100) n/n^+ GaAs layers.
The mirrors were reactive ion etched (RIE) through triangular openings in
MP1350J photoresist at 45^0 to the cleavage planes (see Fig.1). RIE was performed with CCl_2F_2 at 30 mTorr and a power density of 0.4 W/cm^2. The intensity distribution of the waveguide shows that the mode extends into the substrate (Fig.2). Therefore, an etch depth of at least 4 μm is needed for an
epitaxial layer thickness of 2 μm. A second mask consisting of stripes with
different width and a mirror protection was applied and the waveguide ribs
reactive ion etched at the same conditions as above.

An SEM photograph of a waveguide mirror is shown in Fig.3. The (010)
mirror plane is vertical to the surface. The roughness is below 0.1 μm in
magnitude and is believed to be due to mask irregularities. As the photoresist becomes very thin at the top edge of the mirror wall, a small undercut facet (∿0.2 μm high) may result from mask erosion. Changing the succession of the two fabrication steps (rib etch prior to mirror etch) gives
comparable results. In order to insure proper loss measurements, the cleaved input and output facets were antireflection coated with a λ/4 SiO layer.

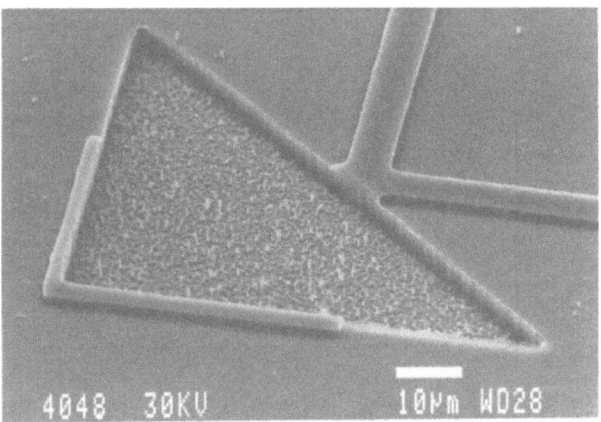

Fig.3 Reactive ion etched rib waveguides with totally reflecting corner
mirror

4 OPTICAL MEASUREMENTS

The transmission of the devices was measured at a wavelength of 1.3 μm by
end-fire coupling of light from a laser diode through microscope objectives.
The output facet was either imaged on a vidicon screen or the output power
measured with a Ge photodetector. The mirror losses were determined by comparing the devices with straight waveguides, and elimination of propagation
and input coupling loss. The loss distribution of single-mode waveguide
mirrors is shown in Table I for different rib width. An average loss of
1.7 dB, with a minimum of 0.4 dB was found for 6.4 and 8 μm wide waveguides.
Higher losses were measured for devices with 4 μm wide waveguides that are
near cut-off of the fundamental mode. The 0.2 μm high undercut facet caused
by mask erosion contributes negligible loss. The calculations show that probably a small mirror displacement is responsible for part of the loss of
these experimental waveguide mirrors.

Table I Measured TE_{00} mirror loss at $\lambda = 1.3$ µm

rib width (µm)	mirror losses (dB)						average loss (dB)
4	4.7	5.1	5.4	5.6	5.6	5.9	5.4
6.4	1.0	1.6	1.8	1.8	1.8	2.0	1.67
8	0.4	1.4	1.5	2.1	2.2	2.3	1.65

A possible application of these mirrors is their integration in a fiber-coupled Y-junction that can be used as a beam splitter at the output of a waveguide device,or as a beam combiner in an integrated phase detector. Only very small branching angles ($<0.5^o$) are allowed for low loss rib waveguide Y-junctions. Therefore,a large chip area is needed to get a distance between the waveguide arms that allows direct coupling to single-mode fibers. The combination of a Y-coupler with corner mirrors avoids this problem (Fig.4). The chip area needed for such a coupler is only 500x50 µm in n/n^+ material and can be reduced substantially by using a heterostructure. First measurements showed mirror losses of 2 to 3 dB, probably due to an insufficient mirror etching depth.

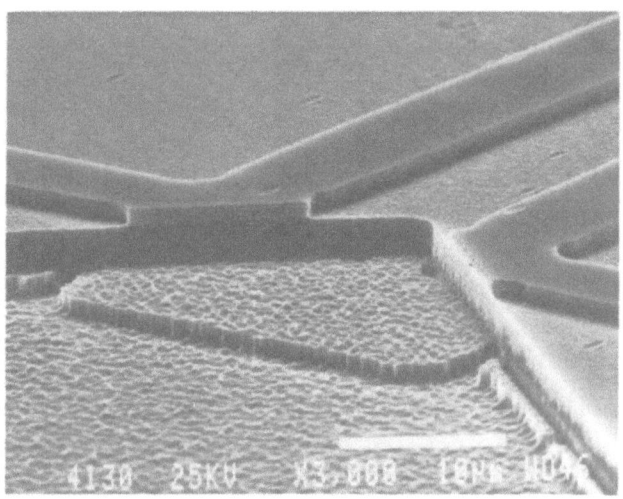

Fig.4 Y-coupler with totally reflecting mirrors at each arm

5 CONCLUSION

The reactive ion etching technique has a number of advantages for both mirror and rib waveguide fabrication in GaAs. The rib profile is independent from orientation on (100) GaAs and waveguide propagation losses down to 2 dB/cm have been measured. Smooth and vertical mirror facets can be fabricated. The method is applicable to angles of incidence other than 45° down to the critical angle for total reflection (18° at $\lambda = 1.3$ µm). It should be possible to reduce the mirror losses by using electron beam lithography instead of emulsion masks. Moreover,even better results are expected for heterojunction waveguides that need smaller etch depths and are less sensitive to non-ideal mirrors.

6 ACKNOWLEDGMENT

This work was supported by the Swiss PTT.

REFERENCES

1 P. Buchmann and A.J.N. Houghton: "Optical Y-junctions and S-bends formed by preferentially etched single-mode rib waveguides in InP", Electron.Lett. 18, 850 (1982)

2 T.M. Benson: "Etched-wall bent-guide structure for integrated optics in the III-V semiconductors", J.of Lightwave Techn. LT-2, 31 (1984)

3 E.L. Hu and R.E. Howard: "Reactive ion etching of GaAs and InP using $CCl_2F_2/Ar/O_2$", Appl.Phys.Lett. 37, 1022 (1980)

4 P. Buchmann, H. Kaufmann, H. Melchior and G. Guekos: "Reactive ion etched GaAs optical waveguide modulators with low loss and high speed", Electron.Lett. 20, 295 (1984)

5 L.A. Coldren, K. Iga, B.I. Miller and J.A. Rentschler:"GaInAsP/InP stripe-geometry laser with a reactive-ion-etched facet", Appl.Phys.Lett. 37, 681 (1980)

6 P. Buchmann and H. Kaufmann: to be published in J. of Lightwave Technology (1985)

Measurement of Semiconductor Optical Waveguide Loss Using a Fabry-Perot Interference Technique

M.W. Austin

Department of Communication and Electronic Engineering, Royal Melbourne Institute of Technology, Melbourne, Australia

P.C. Kemeny

Telecom Research Laboratories, Melbourne, Australia

The propagation loss of semiconductor optical waveguides has been determined by measuring the periodic modulation of transmitted light which occurs when the optical path length is changed.

INTRODUCTION

The measurement of the insertion loss of a planar optical waveguide usually ignores the effect of multiple reflections at air/guide interfaces. For materials with a large refractive index and hence high reflectivity, such as GaAs ($n = 3.44$ at $\lambda = 1.15$ μm), the waveguide forms a Fabry-Perot resonator, and for short guide lengths, neglecting the multiple reflections can lead to a large error in the apparent insertion loss.

We have studied the effect of multiple reflections in rib waveguides fabricated in GaAs/GaAlAs heterostructures grown by molecular beam epitaxy. The transmission of infrared radiation ($\lambda = 1.15$ μm) was measured whilst the guides were heated by 5 - 10°C. Heating produces a change in the optical path length of the waveguide, and this results in a periodic modulation of the transmitted light intensity. Using the properties of a resonator, the observed modulation depth has been used to determine the guide propagation loss.

THEORY

If multiple reflections at the air/guide interfaces of an optical waveguide are considered, then the ratio of transmitted to incident light intensity after propagation through a waveguide of length L and loss coefficient $\propto$ (cm^{-1}) is given by;

$$\frac{I_T}{I_0} = \frac{T_{max}}{(1-R')^2} \cdot \frac{1}{1 + F. \sin^2 \beta L} \tag{1}$$

$$\text{where} \quad F = \frac{4R'}{(1-R')^2} \tag{2}$$

$$\text{and} \quad R' = Re^{-\propto L} \tag{3}$$

T_{max} is the maximum facet transmittance, R is the facet reflectivity and β is the propagaton constant at wavelength λ. It can be seen from equation (1) that the transmittance is a function of βL and thus depends on the optical path length of the waveguide. The shape of the transmittance versus βL curve depends strongly on the value of F [1]. The ratio of minimum to maximum transmitted light intensity, denoted by u, is given by;

$$\frac{I_T (min)}{I_T (max)} = u = \frac{1}{1 + F} \tag{4}$$

Substitution of equations (2) and (3) into equation (4) and rearranging leads to:

$$\alpha L - \ln R = \ln \left[\frac{1 + u^{\frac{1}{2}}}{1 - u^{\frac{1}{2}}} \right]$$

$$= f(u) \tag{5}$$

Thus, a plot of $f(u)$ versus guide length enables the waveguide propagation loss coefficient to be determined.

For a given guide, u may be determined by changing the optical path length and observing the resultant periodic modulation of the transmitted light intensity. The optical path length of a waveguide can be varied by changing the guide length and/or the refractive index, n. Both are temperature-dependent. Defining p as the number of optical half-wavelengths in a guide of length L,

$$p = \frac{2Ln}{\lambda} \tag{6}$$

then the change of p with temperature is given by;

$$\frac{\lambda}{2Ln} \frac{dp}{dT} = \frac{1}{L} \cdot \frac{dL}{dT} + \frac{1}{n} \cdot \frac{dn}{dT} \tag{7}$$

The terms on the right-hand side of equation (7) have numerical values of 6.86×10^{-6} K^{-1} and 6.4×10^{-5} K^{-1} respectively [2,3], and thus the change of refractive index with temperature dominates.

EXPERIMENTAL

The effect of multiple reflections has been studied using rib waveguides fabricated in GaAs/$Ga_{0.94}Al_{0.06}$ As heterostructures grown by molecular beam epitaxy. Optical waveguiding is observed at a wavelength of 1.15 μm by end-fire coupling into the cleaved end of a rib guide via a x 45 microscope objective (NA = 0.65). Light transmitted through the waveguide is detected by a Ge photodiode via another x45 objective. In order to change the optical path length of the guides, they are mounted on an alumina sample holder on which a thick-film resistor has been fabricated. Passing a current through the resistor heats the guides, and therefore changes their length and refractive index. In order to monitor the waveguide temperature, a thermocouple is mounted on a piece of bulk GaAs adjacent to the waveguide sample. The voltage derived from the thermocouple is fed to the x input of an x-y recorder whilst the photodiode output is fed to the y - input. Heating the guides by 5-10°C produces a trace which shows the expected Fabry-Perot oscillations in the detected power. This can be seen in Figure 1. The guide in this case was 2 μm wide and 1.4 mm long. The GaAs guiding layer was 1.9 μm thick and the rib height was 0.37 μm. The waveguide supported only a single lateral mode. Also shown is the theoretical period of oscillation. It can be seen that there is good agreement between the predicted and measured oscillation period. The decrease of transmitted intensity with decreasing temperature is due to a change in coupling conditions. Excessive heating of the sample holder destroys the input coupling and the transmitted light falls to zero.

Measuring the modulation depth of the transmitted intensity enables the function $f(u)$ to be calculated. In Figure 2, $f(u)$ has been plotted for two guide lengths, 1.4 and 3.3 mm. A third point (L = 0) is known since the reflectivity is known (R = 0.3). The slope of the resulting line gives a loss coefficient of 6.0 cm^{-1}. These waveguides are very lossy due to

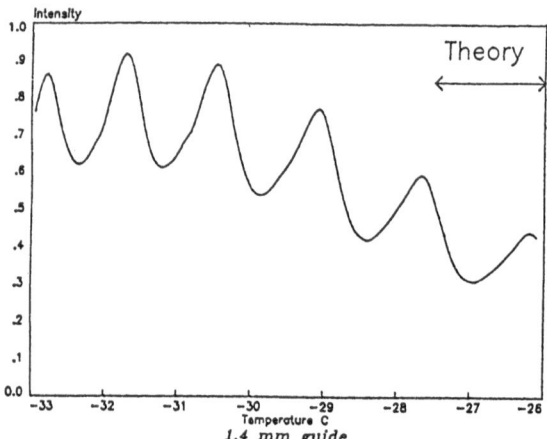

Figure 1. Plot of transmitted light intensity versus waveguide temperature

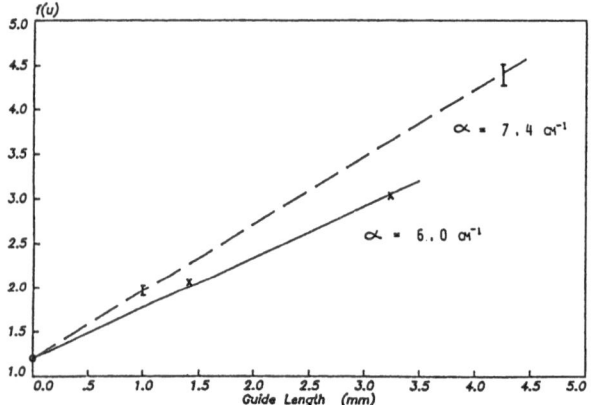

Figure 2. Plot of f(u) versus waveguide length (——) Fabry-Perot measurement; (---) sequential cleaving

scattering from edge defects. This is significant because of the narrow
guide width. Wider guides showed less loss, however they supported two or
three modes, and beating between the modes complicated the plot of transmitted
intensity versus temperature.

Because the transmitted light intensity depends on the optical path length
of the sample, measurements made using the conventional sequential cleaving
technique depend critically upon where the cleave is made. This technique
is therefore susceptible to error. Also shown in Figure 2 are results
derived from sequential cleaving measurements. These give a value for the
loss coefficient of 7.4 cm^{-1}.

CONCLUSIONS

We have reported a novel, non-destructive technique for the measurement of waveguide loss for guides of large refractive index. This technique is more convenient and accurate than the conventional sequential cleaving measurement technique. In applications where the transmission is to be stable with temperature, semiconductor guide-wave devices will require antireflection coatings, not only to improve coupling losses, but also to suppress internal multiple reflections. Alternatively, temperature-dependent transmission could form the basis of a sensitive temperature sensor.

REFERENCES

[1] See for example, M. Born and E. Wolf, 'Principles of Optics' (Pergamon Press, Oxford, 6th ed. 1980) p. 327.

[2] E. D. Pierron, D. L. Parker and J. B. McNeely, Acta Crystallogr 21, 290 (1966).

[3] M. Cardonna, 'Atomic Structure and Properties of Solids, E. Burstein Ed. (Academic Press, New York, 1972) p. 514.

Part IV

Modulators

Parametric Processes in LiNbO$_3$

D.B. Ostrowsky

Laboratoire d'Electro-Optique, Université de Nice, Parc Valrose
F-06034 Nice Cédex, France

1. INTRODUCTION

One of the fields in which integrated optics has yet to realize its full potential is that of optical frequency conversion (second harmonic generation, parametric amplification and oscillation, etc...). While other nonlinear devices such as modulators and switches have attained the performances necessary for their practical utilisation, frequency conversion devices have not.

The purpose of this paper is to examine the reason for this (relative) failure and to determine in view of recent developments, the limits one can hope to attain. We shall begin by discussing, as an example of a parametric process, second harmonic generation (SHG) in general. SHG in lithium niobate in titanium indiffused guides (TI), proton exchanged guides (PE) and guides realised by the combined TIPE process will be examined. We will then go on to discuss the possibilities offered by the TIPE guides for integrated parametric oscillators.

2. SECOND HARMONIC GENERATION: COMPARISON OF BULK AND GUIDED WAVE CONFIGURATIONS

To begin our discussion we shall compare the case of SHG in both bulk and guided wave configurations.

The efficiency of such a process, is defined as $\eta = \dfrac{P_{2\omega}}{P_\omega}$ the ratio between the power generated at the harmonic frequency, $P_{2\omega}$ and the pump power P_ω.

For a plane wave interaction at efficiencies inferior to 20% we have :

$$\eta = d^2 \ \frac{l^2 \ P_\omega}{S} \ \sin^2 \frac{\Delta k l}{2} \ \left(\frac{\Delta k l}{2} \right)^2 \ R$$

where R is an overlap integral, d is the nonlinear coefficient, P_ω represents fundamental optical power, l the interaction length, S the beam cross section and $\Delta k = k(2\omega) - 2k(\omega)$.

The terms $\dfrac{l^2 P_\omega}{S}$ describe the influence of power density confinement over the interaction length and the $\sin^2$ term describes the effect of optical phase matching between the fundamental and the harmonic waves.

Integrated optics provides an important means of optimizing both of these terms. Insofar as the phase matching term is concerned, guided wave optics allows obtention of the phase match ($\Delta k = 0$) via modal dispersion, in cases where the natural crystal birefringence does not permit phase matching. Since phase matching is essential for an efficient interaction this is a major advantage permitting the use, in integrated optics, of materials that cannot be used in the bulk configuration. An important example of this, which we shall discuss later, is the case in which we take advantage of proton exchange to use a particularly strong nonlinear coefficient (d_{33}) in lithium niobate.

However, to continue our comparison, we shall consider cases where phase matching is possible in both the bulk and guided wave configurations In this case the essential advantage of the guided wave arises from the waveguide's ability to confine the optical power density $\frac{P_\omega}{S}$ over long inter-action lengths. In the bulk case diffraction leads to a compromise between a small value of S, and hence a high power density, and a large value of 1.

For gaussian beams the optimum configuration is close to that called confocal focusing where $S = \frac{\lambda 1}{2n}$. If we compare this confocal focusing case to that of guided waves, the ratio of the guided wave term to that of the confocal term is given by :

$$\frac{GW}{CF} = \frac{1}{2n} \frac{\lambda 1}{S} \quad .$$

For n = 2, λ = 1 , l = 2cm and S = $25\mu^2$, values appropriate for LiNbO3, this leads to the rather impressive factor of improvement of about 200 for the guided wave case. However, the efficiency of the guided wave interac-tion is reduced by the <u>overlap integral</u> of the interacting waves, which is proportional to $E(\omega)^2 E(2\omega)$ ds, which in many cases practically cancels the power confinement advantage.

ITO and INABA /1/ recognized the importance of this problem and sug-gested the use of a passive cladding layer to improve the overlap integral We believe appropriate passively clad guides will probably always provide the most efficient configurations for frequency conversion processes. We shall therefore,having reviewed the theoretical background of the sub-ject, examine the experimental situation, to see what has been accomplished to date, and what limits we could hope to attain with the passive cladding technique.

The best results obtained to date in SHG from the near infrared have been realized in Ti : LiNbO3 guides /2/3/4/5/. In table I we resume these results. In the last column, we have given an estimated efficiency norma-lized to the case of a 20 mW pump, l = 2 cm and S = $25\mu^2$. This corresponds to reasonable guide configurations pumped by semiconductor lasers.

The best results obtained correspond to efficiencies around 0.5%, which is about one-half the theoretical efficiency value for these guides. If one deposits cladding layers of Nb_2O_5, for example, on appropriate,.low Ti concentration guides /6/ efficiencies approaching 4% should be attainable. To go beyond this limit we are forced to consider other material systems.

TABLE I. SHG in Ti.LiNbO$_3$

Author	Interaction	Tuning	l(cm)	(4)	P(mω)	η %	(norm)%
UESUGI,KIMURA	TEo - TMo	Temp	1.0	5	20	.015	0.6
UESUGI et al.	TMo - TE$_1$	Temp	2	10	60	.7	0.4
SOHLER, SUCHE	TMo - TE$_1$	λ	1.7	100	45x10^3	25	0.4
ZOLOTOV et al.	TEo - TE$_1$	θ	.8	100	?	16%	0.4

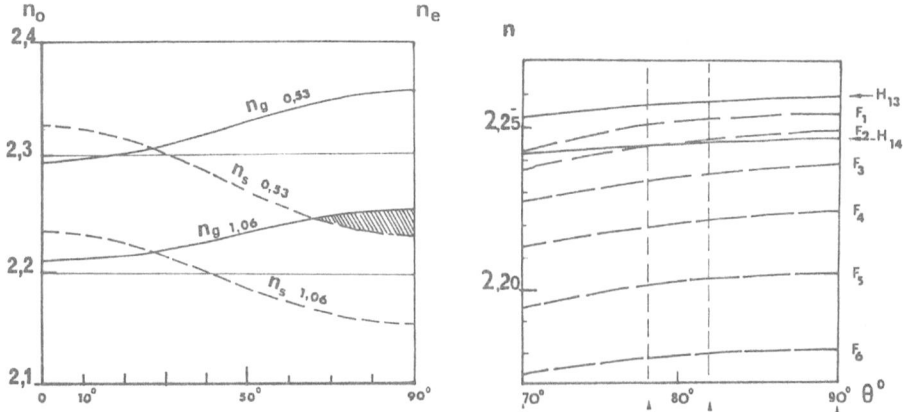

Figure 1. Substrate and surface indices of TE waves at 1.06 and 0.53 microns in a proton exchanged X-cut lithium niobate guide as a function of propagation angle with respect to the optical axis.

Figure 2. Effective indices of the fundamental modes (TE_1-TE_6) and harmonic modes (TE_{13} and TE_{14}) as a function of the angle of propagation with respect to the optical axis. These values were measured at 70°, 78°, 82° and 90°.

3. PROTON EXCHANGED GUIDES

3.1 Second harmonic Generation

As an example of what one can hope to attain with other systems, we shall describe the use of PE guides, which permit realising phase matching with the d_{33} non-linear coefficient in lithium niobate.

The use of the d_{33} coefficient is of considerable interest since it is approximately 6 times the commonly used d_{13} coefficient and could lead, therefore, to conversion efficiencies on the order of 36 times those previously attained.

The conversions observed were, however, several orders of magnitude smaller than those expected, possibly due to a corresponding reduction of the LiNbO₃ non-linearity due to the proton exchange.

In SHG experiments performed to date in lithium niobate, phase matching is accomplished by using orthogonal polarizations for the fundamental and harmonic waves (coupled by the d_{13} coefficient) permitting wavelength dispersion compensation by birefringence. In the experiment, we describe here /6/ the important change in n_e ($\Delta n_e \sim 0.12$) induced by proton exchange /7/ allows a modal compensation of the material dispersion permitting the use of fundamental and harmonic waves having the same polarization (coupled by the d_{33} coefficient).

This can be understood by regarding figure 1. We have plotted for an X-cut crystal the substrate and surface indices of a TE wave at 1.06 µm and 0.53 µm of proton exchanged lithium niobate as a function of θ, the angle between the direction of propagation and the optical axis. In the cross hatched region the surface index at 1.06 µm is superior to the substrate index at 0.53 µm. This allows the compensation of wavelength dispersion, albeit for low order fundamental modes coupled to high order harmonic modes with tuning allowed by a variation of θ.

The samples were prepared by proton éxchange in benzoic acid for 20 hours at 200 °C.

The samples were then characterized via prism coupling at 1.06 and 0.53 microns. Figure 2 shows the effective mode indices measured for the first 6 fundamental modes and the 13th and 14th harmonic mode as a function of θ. Phase match is seen to occur at $\sim 78°$ for a TE_2 (1.06) TE_{13} (0.53) interaction (having an effective overlap integral of $\sim 10^{-4}$).

This prediction was confirmed experimentally. A Q-switched Nd : Yag laser was prism coupled into the guide and the guide rotated to the appropriate angle. At this angle ($\sim 78°$), with a fundamental power of 2.4×10^3 W, having a beam-width of 1 mm, recoupled from the output prism a peak harmonic power of 1,2 W was observed. This is on the order of 1.5×10^{-3} times the SHG power expected.

These results appear to corroborate at optical frequencies the results of BECKER /8/which suggested a reduction of the low frequency electrooptic effect in proton exchanged guides.

Despite this rather disappointing result, we do not believe that this is a closed subject. Research on proton exchange is actually in its infancy and there are reasons to believe that it will be possible to realize PE or TIPE guides which maintain the necessary non-linearity. The TIPE guides, in particular have already been used to extend the SHG range in lithium niobate and, at low proton concentration, do not appear to affect the non-linear coefficient /9/.

3.2 Parametric oscillation in TIPE Guides

Let us now go on to consider what one can hope to attain using TIPE guides for parametric oscillators.

In particular we shall show that it should be possible to fabricate parametric oscillators, pumped by GaAs lasers, tunable over the 1.3 and 1.55 μm ranges. Evidently these regions, where fiber dispersion can be minimized, are of major interest for frequency multiplexed and coherent transmission systems, which would be ideal "clients" for tunable parametric oscillators.

In bulk or titanium doped lithium niobate, parametric oscillators functioning around 1.3 or 1.55 μm must be pumped with visible sources (typically dye lasers). To pump such oscillators with diode lasers ($\lambda \sim 0.85$ μm) it would be necessary to heat the crystals to around 500 °C, at which point the crystals degrade due to desoxygenation.

As we shall show, the use of TIPE guides should permit us to circumvent this problem.

In a parametric oscillator, three waves, called the pump, signal, and idler waves, interact. A pump photon is converted into a signal and a an idler photon. Energy conservation implies :

$$\frac{1}{\lambda_p} = \frac{1}{\lambda_s} + \frac{1}{\lambda_i}$$

where λ_p, λ_s and λ_i are the pump, signal and idler wavelengths, respectively.

To satisfy the phase matching condition we must also satisfy :

$$\frac{n_p}{\lambda_p} = \frac{n_s}{\lambda_s} + \frac{n_i}{\lambda_i}$$

where n_p, n_s and n_i are the pump, signal, and idler indices respectively.

When dealing with guided waves we simply replace these bulk indices by the effective indices of refraction of the interacting modes.

Note that since a change in temperature changes the indices, and the phase match condition, it is possible to tune the signal and idler frequencies while maintaining the pump frequency fixed.

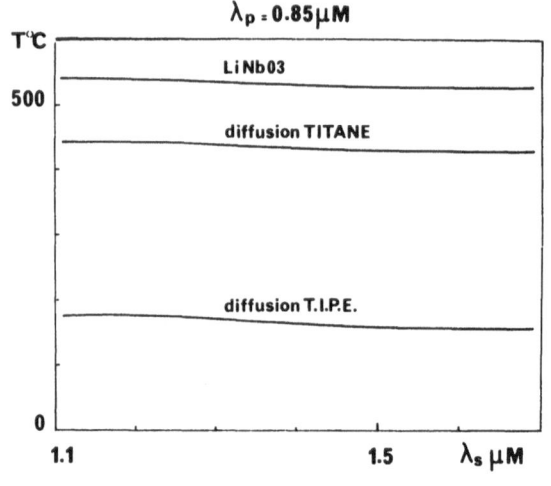

Figure 3. Phase matching temperature for LiNbO$_3$ and Ti:LiNbO$_3$ and TIPE guides with 0.85 µm pump.

Figure 4. Parametric phase match diagrams at various temperatures as a function of pump wavelength.

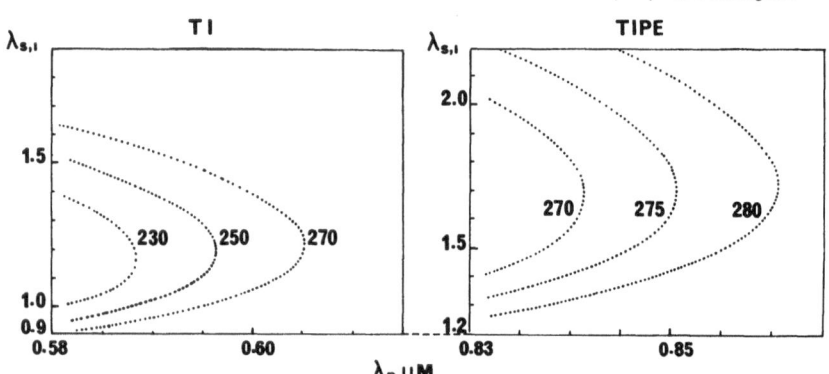

In fig.(3) we show the phase matched signal wavelength, λ_s as a function of temperature, for bulk lithium niobate, TI guides and TIPE guides pumped by a 0.85 µm source using a TE$_0$-TM$_0$-TM$_0$ interaction.

The important reduction of the operating temperature for TIPE guides is evident.

A comparison of TI and TIPE guide parametric phase match is shown in fig.(4) as a function of pump wavelength.

There we note how, at reasonable operating temperature, semiconductor pump wavelengths become possible.

4. CONCLUSION

In conclusion, it seems clear that integrated optical interactions in TI, PE, and TIPE guides, offer tantalizing possibilities for the realization of tunable sources throughout the visible and near infrared regions pumped by diode lasers.

Whether we shall overcome the material and microstructure problems involved depends, I believe, simply on the level of effort we are willing to provide

While the responsibility for this view is my own I would like to acknowledge many fruitful discussions and assistance provided by my colleagues, M. Papuchon and M. De Micheli.

REFERENCES

1. H. ITO and H. INABA, Opt. Letts. 2 139 (1978)

2. W. SOHLER and H. SUCHE, Appl. Phys. Letts.33 518 (1978)

3. N. UESUGI and T. KIMURA, Appl. Phys. Letts. 29 572 (1976)

4. N. UESUGI, K. DAIKOKU and M. FUKUMA, Appl. Phys. Letts. 49 4945 (1978)

5. E.M. ZOLOTOV, V.M. PELEKHATYI, A.M. PROKHOROV and V.A. CERNYKH, JETP 49 603 (1979)

6. W. SOHLER and H. SUCHE, Private Communication

7. J.L. JACKEL, C.E. RICE and J.J. VESELKA, Appl. Phys. Letts. 41 607 (1982)

8. R.A. BECKER, Appl. Phys. Letts. 43, 131 (1983)

9. M. DE MICHELI, J. BOTINEAU, S. NEVEU, P. SIBILLOT and D.B. OSTROWSKY, M. PAPUCHON, Appl. Phys. Letts. 8 116 (1983)

Electrooptic Modulators in Multilayered Zinc-Oxyde Waveguides

Winfried H.G. Horsthuis and Reinier Pannekoek

Twente University of Technology, Department of Electrical Engineering
P.O. Box 217, NL-7500 AE Enschede, The Netherlands

A novel type electrooptic mode converter in a multilayered
zinc-oxyde waveguide is reported. The intended application is
voltage or electric field sensing.

Introduction

Electrooptic modulation in integrated optics is of great importance for applications in optical communications and optical
sensors. We investigate modulators in zinc-oxyde planar waveguides for application in integrated optic sensors for voltage
or electric field strength. To improve accuracy a reference
signal is very important in intensity modulating sensors. We
therefore are especially interested in modulators based on
mode-coupling phenomena. These modulators intrinsically provide
us with such a reference signal, since the output consists of
two separate signals: the originally excited beam and the
coupled beam. In this context Bragg deflectors may be regarded
as mode-couplers as well. Here coupling takes place between
modes in the continuum of possible directions in plane of the
slab itself.

Two types of modulators are investigated: A Bragg deflector
and a novel type mode-coupler which couples the two guided
modes of the same polarization in a multilayered waveguide.
Such a waveguide has a specific mode structure with symmetric
and antisymmetric modes having only small differences in propagation velocities.

Mode coupling in multilayered waveguides

The majority of reported mode couplers describe guided TE-TM
coupling [1-3]. The required spatial period to match the difference in propagation constants of the two modes is given by:

$$\Lambda = \lambda_o / \Delta N_{eff} \tag{1}$$

where λ_o is the vacuum wavelength of the light and ΔN_{eff} the
difference in effective mode indices. In order to couple two
modes of the same polarization Λ becomes very small because of
the relatively large ΔN_{eff} of those modes. For a ZnO film
which supports a few modes Λ is about 3 μm. Only in case of
rather thick waveguides with many modes (> 1 μm) Λ increases to
about 10 μm. A disadvantage in thick slabs is the increased
loss, the presence of many modes and the drop in electric field
strength for perpendicular fields.

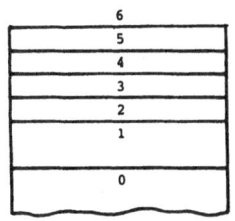

6: air
5: 0.30 μm SiO_2(C.V.D.)
4: 0.23 μm ZnO
3: 0.20 μm SiO_2(C.V.D.)
2: 0.25 μm ZnO
1: 1.00 μm SiO_2(thermal)
0: silicon

Fig. 1. Cross-section of multilayered waveguide

Another possibility for increasing the required spatial period is the use of a multilayered waveguide.
Figure 1 shows a cross-section of such a waveguiding system.
These multilayered waveguides require the use of deposition techniques, since by diffusion these systems cannot be made. The large difference in refractive indices of ZnO and SiO_2 (2 resp. 1.46) is another essential point. Due to this large difference, the penetration depth of the optical fields is very small. We therefore only need thin buffers of SiO_2. Figure 2 shows two calculated examples of the mode structure of multi-layered waveguides. In fig.2a the two ZnO cores have equal thicknesses, and have a separation of 0.3 μm. The symmetric and antisymmetric modes nearly have the same effective indices, 1.83007 and 1.82305 respectively. The system of fig.2b has 10% difference in core thickness and a 0.2 μm separating layer. The two effective indices are 1.83211 and 1.80422 for the symmetric and antisymmetric modes respectively. It is obvious that the period Λ for coupling these modes is a strong function of the ratio of the two ZnO-core thicknesses. Standard photolithographic processes restrict this period to be larger than 6 μm. On the other hand, this period should not be too large, since the required interaction length to obtain the same modulation depth at constant voltage linearly depends on this period.

Experiments.

We carried out some experiments on mode coupling in multi-layers. The fabricated structure was intended to have the lay-

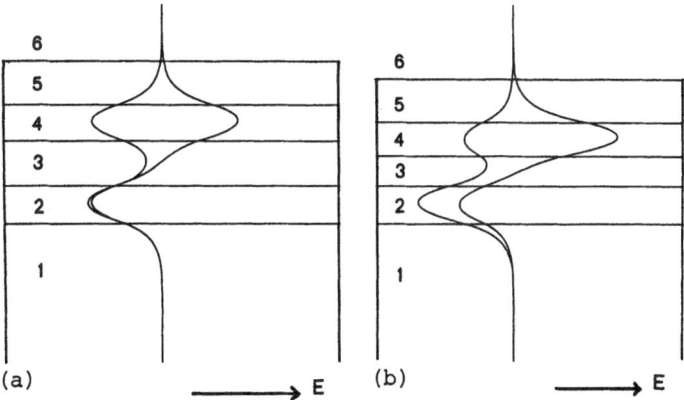

(a)   E   (b)   E

Fig. 2. Mode structures in multilayers

out depicted in fig.1, consequently having the mode structure of fig. 2b.

A silicon wafer of orientation 1-1-1 was steam oxidized for 3 hours at 1150 °C. The first ZnO layer was sputtered at 1800 W for 75 seconds. Then SiO$_2$ was deposited (C.V.D.), and the next ZnO film was sputtered for 69 seconds at the same RF power. Finally the SiO$_2$ cladding was deposited. It should be noted that no thickness monitors were available, so a stopwatch determined the thicknesses of our layers.

The mode indices were determined by measuring the coupling angles from a rutile prism and subsequent calculations. We found 1.8266 $\pm$0.0004 for the symmetric mode, and 1.7914 $\pm$0.0004 for the antisymmetric mode. With (1) this results in a required coupling period of 18 $\pm$0.5 μm.

With a lift-off technique an aluminum electrode pattern with period s=16 μm is placed on top of the waveguide. The interaction length is 1 mm, or 62 periods. The back-side of the wafer was metallized to form the ground electrode. Because of the very low resistivity of the wafer,the actual electrode is only separated from the waveguide by the thermal oxide layer. With this electrode configuration homogeneous perpendicular fields can be generated in the waveguide.

We first determined the required coupling periodicity by rotating the wafer with respect to the light beam. The periodicity can be calculated from Λ = s/sinα , with α the angle between electrodes and light beam. We measured α=50 $\pm$3 degrees, resulting in a period Λ = 18.8 $\pm$0.5 μm, in agreement with the calculated value. Coupling already occurs in the absence of an electric field,since the evanescent fields interact with the electrodes. The measured coupling without an applied field was approximately 40%.

Figure 3 shows two oscilloscope pictures of the voltage-induced modulation. In fig. 3a the driving voltage is 30 V eff. The intensity of the symmetric mode follows the modulating signal, while the antisymmetric mode was launched with the prism-coupler. For -65 V input voltage the symmetric mode is completely depleted. For even lower voltages light is again coupled into this mode, which is clearly seen in fig.3b where the driving voltage is increased to 90 V eff. From these measure-

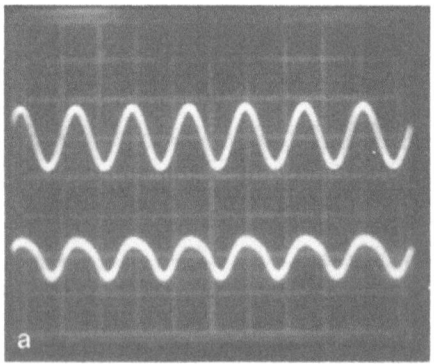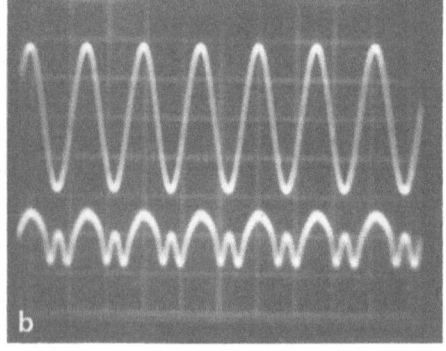

Fig. 3. Lower trace: TE-TE mode converter respons.
       Upper trace: driving voltage.

ments, it is possible to check several waveguide and material parameters, since apparently -65 V cancels the phase-shift due to the presence of a metal layer in the vicinity of the waveguide.

We calculated the change in effective mode index for slab waveguides with and without metal claddings. For a waveguide of 0.3 μm thickness, and a buffer of 0.3 μm SiO_2 this change in effective mode index is approximately 10^{-4}.

We now can calculate the change in refractive index due to the applied field, according to:

$$\Delta n = \frac{1}{2} n_o^3 r_{13} E_x \tag{2}$$

With $n_o = 1.983$, $r_{13} = 1.4 \cdot 10^{-12}$ m/V and $E_x = 18 \cdot 10^6$ V/m (wich results from a voltage of 65 V) we find $\Delta n \approx 10^{-4}$, in agreement with the expected value.

We expect to improve the sensitivity of this device by at least one order of magnitude. This can be realized by increasing the interaction length, and optimizing the electrode configuration.

Bragg modulation

The second type of modulator we investigated is a Bragg-deflector. Bragg deflection is governed [4,5] by the next equations:

$$I = I_o \sin^2(\frac{1}{2}(\Delta\phi + \phi_o)) \tag{3}$$

where I and I_o are the intensities of the deflected beam and input beam respectively, ϕ_o is the bias optical phase retardation due to the electrodes near the waveguide, and $\Delta\phi$ is the induced phase-shift due to the electrooptic effect given by:

$$\Delta\phi = (- 2\pi L/\lambda_o) n_o^3 r_{13} E_x \tag{4}$$

with L the interaction length, r_{13} the appropriate electro-optic coefficient for TE modes and E_x the field strength perpendicular to the slab.

The Bragg angles in the guide and in air are given by [4]:

$$\sin\theta_B = \pi/2\Lambda\beta \tag{5}$$

and

$$\sin\theta_B = \lambda_o/2\Lambda \tag{6}$$

respectively. Here Λ is the period of the grating, and β is the propagation constant of the mode under investigation.

Experiments

Several deflectors have been realized. They consisted of a 0.6 μm thick ZnO layer on a thermally oxidized silicon wafer. An isolating buffer of 0.2 μm SiO_2 separates the aluminum electrodes from the waveguide. The top electrodes have a periodicity of 10 μm. Again, the metallized substrate forms the ground electrode.

Several interaction lengths are fabricated. The intensity of the bias deflection is dependent on the thickness of the SiO_2 buffer and the interaction length. We measured the deflected beam intensity as a function of the applied field at several

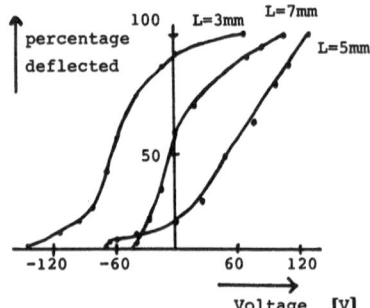

Fig. 4. Results of Bragg deflection measurements

interaction lengths (L) at 1 KHz. The results are shown in figure 4. From the figure we conclude that the sensitivity of the
modulator increases with increasing interaction length. Full
modulation occurs for 130 V and 7 mm interaction length. The
bias phase retardation appears not to be linear dependent on
the interaction length. The cause of this unexpected behaviour
is under investigation. The Bragg angles are measured in the
guide and in air. In table 1 these values are compared with the
calculated ones.
    We also measured strong deflection for an angle of 3.73°
in the guide. This corresponds with a periodicity according to
(5) of about 2.5 μm, which is the third harmonic spatial wavelength of the electrodes.

Table 1. Calculated and measured Bragg angles

	guide	air
calculated	0.92	1.81
measured	0.89 +0.05	1.88 +0.08

Conclusions

We have shown that efficient electrooptic modulation in zinc-
oxyde waveguides is possible. The efficiency of the novel guided TE-TE mode coupler exceeds that of a Bragg deflector with
the same interaction length.
    If these structures could be made with materials with larger
electrooptic coefficients such as LiNbO$_3$, very low driving
voltages (<1 V) would be needed for complete power transfer.
The use of conductive substrates gives the opportunity to cre-
ate homogeneous perpendicular fields in the waveguides. We consequently obtain optimal overlap integrals between optical
fields and applied electric fields.

Acknowledgements.

The authors wish to thank Harry Kreuwel for computer simulations. These investigations were supported by the Netherlands
Technology Foundation (STW).

References

1. D.Marcuse, IEEE J. of Quantum Electr., 11, 759, (1975).
2. R.C.Alferness e.a., Appl. Phys. Lett., 39, 131, (1981)
3. J.F.Lotspeich, J. of Lightwave Techn., 2, 694, (1984)
4. J.M.Hammer e.a., Appl. Phys. Lett., 23, 176, (1973)
5. Y.K.Lee e.a., Appl. Opt., 15, 1565, (1976)

Electro-Optic In-Line Frequency Translator: Performance Limitations

F. Heismann and R. Ulrich

Technische Universität Hamburg-Harburg, D-2100 Hamburg 90, Fed. Rep.of Germany

The frequency translator with interdigital electrodes is analyzed by Fourier expansion in terms of moving grating components. Perturbation theory yields carrier suppression, spurious sideband levels caused by parasitic fields, and driving tolerances.

The in-line frequency translator, employing electro-optic TE → TM polarization conversion, is analyzed. The interdigital electrode pattern and the applied voltage distribution are Fourier-expanded into a series of moving grating components. Perturbation theory yields the levels of residual carrier (due to parasitic field components), image side band, and operating tolerances.

In previous publications [1,2] we had described the concept and realization of an integrated electro-optic frequency translator. It is based on collinear forward Bragg-scattering at a travelling grating of index perturbations, driven by two phase quadrature voltages, see Fig.1. This concept promised high efficiency, good suppression of the residual carrier by an output polarization filter, and low spurious modulation sidebands. In our realizations, however, the carrier suppression was limited to only 30 dB, far below the values of ≥ 60 dB which are desirable for the use, e.g. in a fiber-optic gyro. Similar problems existed with the other spurious sidebands. We have established now that the layout of the interdigital electrode pattern (Fig.2)has a crucial influence on the attainable level of undesired modulation products (Fig.3). In the following, therefore, we outline a simple perturbational analysis of this frequency translator, showing the relations between the electrode pattern, the applied voltages, and the sideband structure.

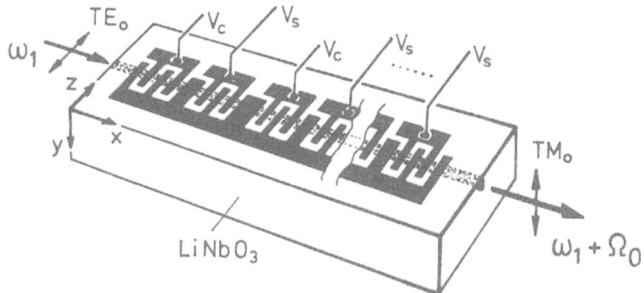

Fig.1. Schematic arrangement of electro-optic frequency translator. The electrode pattern is shown simplified here.

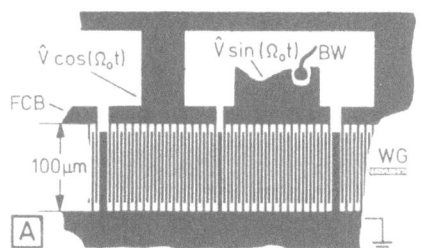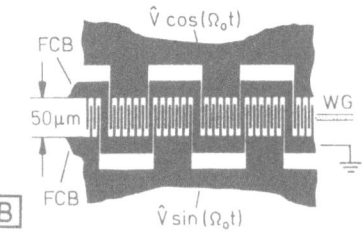

Fig.2. Tested electrode patterns of 2 MN finger pairs. A: M=15 N=16.
B: M=16 N=40. (WG waveguide, BW bondwire, FCB finger-connecting bar).

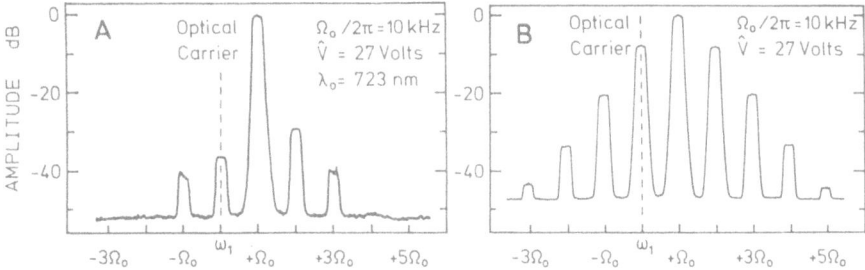

Fig.3. Optical output spectra of the frequency translator, obtained
with the electrode patterns A and B of Fig.2.

The translator operates with a stripe waveguide, supporting two orthogo-
nally polarized modes of amplitudes $a_1(x) \cdot \exp(i\beta_1 x - i\omega_1 t)$ and
$a_2(x) \cdot \exp(i\beta_2 x - i\omega_2 t)$, with propagation constants β_1 and β_2, respectively.
We assume arbitrarily mode 1 to have quasi-TE polarization, and mode 2
quasi-TM. At the input (x=0), only mode 1 is excited, at the output (x=L),
only mode 2 is analyzed. When the modes are coupled by a single discrete
polarization-converting perturbation, the optical phase of $a_2(L)$ depends
linearly upon the position of that perturbation. When the perturbation is
moved with velocity v along the guide, a phase shift of 2π is produced for
every movement through one beat length, $L_p = 2\pi/|\beta_2 - \beta_1|$, of the two modes,
and consequently a frequency shift   of $\Delta\omega = (\beta_2 - \beta_1)v$ results.

In practice, satisfactory coupling efficiency is achieved only with a
periodic pattern of perturbations (Fig.1). Its period length Λ should be
phase matched to the beat length L_p. Such polarization-coupling perturbations
are induced electro-optically [3] in the waveguide material by a system of
interdigital electrodes. By driving different sections of the electrodes
(Fig. 2) with phase quadrature voltages $V_c = \hat{V}\cos\Omega_0 t$ and $V_s = \hat{V}\sin\Omega_0 t$, the
perturbations are moved along the guide in a similar fashion as the fields
of a linear induction motor move along the pole pieces. The resulting opti-
cal frequency shift is equal to the electrical driving frequency, $|\Delta\omega| = \Omega_0$.

We have fabricated and tested a number of such devices with the following
common features: Ti-diffused, x-propagating guides on y-cut LiNbO3, operating wavelength $\lambda_0 = 2\pi c/\omega_1 = 723$ nm, beat length $L_p = 9$ µm, SiO2 buffer
layer of 150 nm, Aℓ electrodes, precision electronic drive voltages of
$\hat{V} \approx 27$ volts peak amplitude, $\Omega_0/2\pi \approx 1 - 10$ kHz, tunable dye laser, lens-
coupling through polished end faces, homodyne detection of all modulation

159

products, using an 80 MHz bulk Bragg cell and an RF spectrum analyzer. Typical output spectra, obtained with electrode patterns A and B of Fig.2 are shown in Fig.3. In both patterns the finger electrodes are connected in groups. In pattern A, all cos-groups are interconnected directly by the metallization, whereas the sin-groups are interconnected with bond wires. In that respect, pattern B is simpler. It uses a meandering ground electrode and requires no bond wires. The output spectrum of B shows much higher levels of undesired modulation products, however.

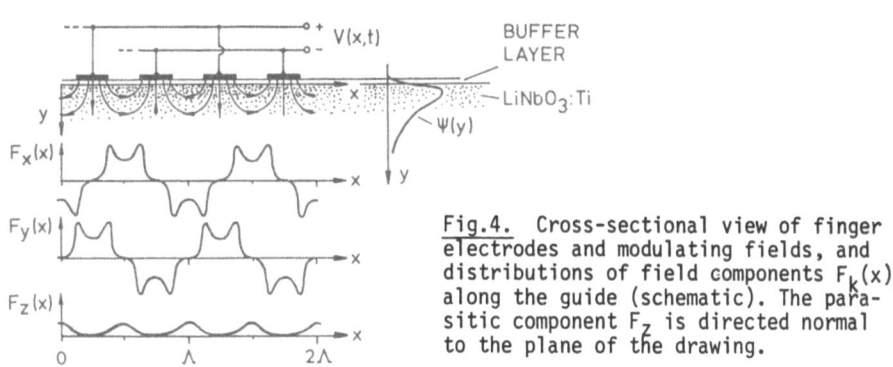

Fig.4. Cross-sectional view of finger electrodes and modulating fields, and distributions of field components $F_k(x)$ along the guide (schematic). The parasitic component F_z is directed normal to the plane of the drawing.

Theory. The distribution of the modulating electric field around the finger electrodes is sketched in Fig.4. All finger pairs within a group are driven with the same voltage $V(x,t)$, which is equal either to $V_c(t)$ or $V_s(t)$, depending on the location x. Therefore, the modulating field strength can be expressed as $\vec{E} = V\vec{F}/\Lambda$, by using here the period length Λ to define a normalized field distribution $\vec{F}(x,y,z)$. The resulting perturbation of the dielectric tensor near the guide is $\tilde{\varepsilon}_{ij} = (V/\Lambda) \Sigma r_{ijk} F_k$, where r_{ijk} are the components of the electro-optic tensor and F_k those of $\vec{F}$. A standard coupled mode analysis yields

$$da_1/dx = i\, c_{11}\, a_1 + i\, c_{12}\, a_2 \exp(i\beta x - i\Omega t) \tag{1}$$

$$da_2/dx = i\, c_{22}\, a_2 + i\, c_{21}\, a_1 \exp(-i\beta x + i\Omega t) \tag{2}$$

$$\kappa_{\mu\nu}(x) = \beta_\mu \beta_\nu^2 c^2 (2\Lambda\omega_\mu^2)^{-1} \sum_{i=1}^{2} \sum_{j=1}^{2} \sum_{k=1}^{3} r_{ijk}\, J_{ijk}^{(\mu\nu)}(x) \tag{3}$$

$$J_{ijk}^{(\mu\nu)}(x) = \iint \vec{\Psi}_{\mu i}^{*}(y,z)\, F_k(x,y,z)\, \vec{\Psi}_{\nu j}(y,z)\, dy\,dz \tag{4}$$

Here, $\vec{\Psi}_1$ and $\vec{\Psi}_2$ denote the vectorial modal fields of TE and TM polarization, respectively, normalized by $\iint \vec{\Psi}_\mu^* \cdot \vec{\Psi}_\nu\, dy\,dz = \delta_{\mu\nu}$, so that $|a_\mu|^2$ is the power in mode μ. The $J(x)$ are their local overlap integrals with the modulating field $\vec{F}$. Several of these $J(x)$ are weighted with the r_{ijk} and combined into the four *normalized* coupling functions $\kappa_{\mu\nu}(x)$. Furthermore, the *actual* coupling coefficients are $c_{\mu\nu}(x,t) = \kappa_{\mu\nu}(x) \cdot V(x,t)$. This product of $\kappa_{\mu\nu}$ and V corresponds to the practical situation, with $\kappa_{\mu\nu}$ given by the finger geometry (see Fig.4), and V determined by their interconnections and the applied voltage distribution. We also abbreviated $\beta = \beta_2 - \beta_1$ and $\Omega = \omega_2 - \omega_1$.

When the driving voltages have frequency Ω_0, the output wave $a_2(L)\exp(-i\omega_2 t)$ must have a line spectrum, containing possibly a residual carrier at ω_1 and a number of sidebands at $(\omega_1 + m\Omega_0)$ with $m = \pm 1,2,3 \dots$. We calculate the amplitude U_m of the m-th sideband by formal integration of (2) over $x = 0\dots L$ and over a sufficiently long time interval, $T \gg 1/\Omega_0$.

$$U_m = iT^{-1} \int_0^T \int_0^L a_1(x,t)\, V(x,t)\, \kappa_{21}(x)\, \exp(-i\beta x + im\Omega_0 t)dxdt \qquad (5)$$

$$+ iT^{-1} \int_0^T \int_0^L a_2(x,t)\, V(x,t)\, \kappa_{22}(x)\, \exp(-i\Omega t + im\Omega_0 t)\, dxdt$$

Although the functions $a_1(x)$ and $a_2(x)$ are not known (they depend on $V\kappa_{\mu\nu}$), this equation permits a good qualitative and quantitative understanding of the translator. We ignore temporarily the 'self-coupling' effect described by κ_{22} in the second integral. We may view then the product $V\kappa_{\mu\nu}$ as the superposition of a number of sinusoidal grating components of the $\tilde{\varepsilon}$ tensor, moving with various propagation factors $\exp(iKx - i\Omega t)$ along the guide. The first integral in (5) is interpreted then as a generalized Fourier integral, expressing $U_m \approx i\bar{a}_1(V\kappa)_m L$ by the amplitude $(V\kappa)_m$ of that moving grating component whose spatial frequency is phase matched to the beat length, $K = \beta$, and whose temporal frequency is $\Omega = m\Omega_0$. All other moving grating components average out to small spurious levels in (5). The quantity $\bar{a}_1$ is the mean value of $a_1(x)$. In situations of large coupling efficiency $a_1 \to a_2$ we may approximate $a_1(x)$ and $a_2(x)$ by the solutions of an ideally phase matched, uniform coupler, $\hat{a}_1 = \cos V_0 h_{12}x$ and $\hat{a}_2 = i \sin V_0 h_{12}x$, with $V_0 h_{12}$ representing the relevant Fourier amplitude. With these functions $\hat{a}_1$ and $\hat{a}_2$, equation (5) assumes the meaning of a first order perturbation result, valid for translators with good coupling efficiency. In particular, $\bar{a}_1 = \bar{a}_2 = 2/\pi$.

Our electrode patterns (Fig.2) have the general arrangement of fingers $[(\bar{c}oco)^M\bar{s}(\bar{s}oso)^M\bar{c}\bar{c}\bar{c}]^N$. The symbols c and s stand for fingers connected to V_c and V_s, respectively, while $\bar{c}$ and $\bar{s}$ are ground electrodes, and o denotes a space. Each of these elements occupies a length $\Lambda/4$ along the x-direction. Groups of M successive c or s fingers are interconnected by *finger-connecting bars* (FCB in Fig.2), and the resulting double group is repeated N times. The voltage function can be expressed as

$$V(x,t) = R(x)\, \hat{V} \cos\Omega_0 t + [1 - R(x)]\hat{V} \sin\Omega_0 t \qquad (6)$$

where $R(x)$ is a periodic 'rectangular' function with period length $\ell = (2M+1)\Lambda$, and $R=1$ when $0 < x < \ell/2$, and $R=0$ when $\ell/2 < x < \ell$. The peak voltages applied are $\hat{V}$. The normalized coupling functions $\kappa_{\mu\nu}(x)$ are periodic within each group with spatial frequency $K_0 = 2\pi/\Lambda$, because all contributing fields have at least that periodicity, see Fig.4. As adjacent groups are offset by $\pm\Lambda/4$ from a uniform periodicity, we have the general form $\kappa_{\mu\nu} = \bar{h}_{\mu\nu} + h_{\mu\nu}(R+iR-i)\exp(iK_0 x)$, ignoring higher terms. Combining with (6) and Fourier transforming yields $\kappa_{\mu\nu}V$ as the described superposition of moving gratings with propagation factors $\exp[iK_0 x + is\Gamma x \pm i\Omega_0 t]$ with $\Gamma = 2\pi/\ell$ and integer s.

The strongest moving grating component in $V\kappa_{21}$ is $(h_{21}\hat{V}/2)\exp(iK_0 x - i\Omega_0 t)$. In the first integral of (5) it produces only the desired up-shifted sideband, $m = +1$. It is phase matched when the optical wavelength is tuned to $\lambda_0 = 723$nm so that $\beta = K_0$. The carrier and image sideband are not produced by the first integral in (5) in that situation. At other optical wavelengths, $\lambda_s \approx \lambda_0[1 + s/(2M + 1)]$, with $s = \pm 1, \pm 3, \pm 5 \dots$, however, weaker grating

161

components become phase matched, and frequency-translating coupling should be expected. With electrode pattern A we have indeed observed large coupling efficiencies (m=-1) at λ_1=742 nm and λ_{-1}=704 nm, but could not obtain coupling for s = ±2.

The second integral in (5) produces a residual carrier U_0 and spurious sideband U_2 by the action of the 'parasitic' component F_3 of the modulating field. This F_3 represents the field between the FCB electrodes. It is fairly weak between the fingers. It increases rapidly with distance $|y|$, and is spatially non-alternating within each group of fingers. Therefore, the $J_{1}(22)$ overlap integral contributes through κ_{22} to U_0 and U_2. Expressed differently, the uniform part of F_3 acts in the fashion of a simple phase modulator, producing a full PM spectrum of sidebands of the up-shifted wave a_2. An equivalent modulation by F_3 occurs also with the input wave a_1, but shows up here only in *second* order perturbation. We believe these modulations to cause most of the observed spurious sidebands in Fig.3. For a quantitative estimate, we have measured $F_3(x)$ with an electrode model in an electrolytic tank, obtaining mean values $<F_3> \approx 10^{-3}$ for pattern A and $<F_3> \approx 4.3 \cdot 10^{-3}$ for B. For $\hat{V}$ = 27 volts we calculate U_0 levels of -26dB and -13dB for A and B, respectively, in reasonable agreement with the measured values of -30dB and -8dB. We recognize that the poor U_0 of pattern B is caused by its short finger length, causing a large $<F_3>$. For a substantial reduction of U_0 it appears necessary to employ either a balanced electrode pattern with $<F_3>$ = 0, or to shield the guide region from F_3 by a metallic ground plane.

The optical bandwidth of the frequency translator can be found from a straightforward strict solution of (1), (2), assuming an ideal, uniform coupler. The resulting 'power bandwidth' can be expressed as $\Delta_* \approx 1.25/L$, where $\Delta=(\beta-K_0)/2$ represents the phase mismatch from detuning. At mismatches of $\pm\Delta_*$ the output power $|a_2|^2$ drops by 3dB. In the wavelength scale, this bandwidth is $\delta\lambda \approx \pm 4.2 \lambda^2/L$, using the group birefringence $(n_{0,gr} - n_{e,gr}) \approx 0.095$ of LiNbO$_3$. For our translators we calculate a full bandwidth $2\delta\lambda = 0.98$ nm, in perfect agreement with the measured 1 nm. In applications, however, where the level of undesired sidebands is critical, another definition of bandwidth may be relevant. Ignoring here the mentioned phase modulation (κ_{22}=0), the image sideband (m=-1) vanishes when $\Delta\to0$. It reappears, however, when the wavelength is slightly detuned, $|U_{-1}(\Delta)/U_1(0)| \approx L\Delta/4N$. This follows directly by evaluating (5). Obviously, for a good suppression of the image sideband, the number N of (s+c) groups in the electrode pattern should be chosen as large as possible (keeping the total length L=(2M+1)NΛ constant). The best choice is N=L/Λ but it requires the most complicated and longest electrode pattern. In our coupler B with N=40 the full 'purity bandwidth' for 40dB image sideband suppression is $2\delta\lambda \approx 0.62$ nm, slightly less than the 'power bandwidth'.

Finally we can use the perturbation expression (5) to determine tolerances for the permissible deviations in the amplitudes and in the phase difference of the two quadrature voltages. An antisymmetric amplitude deviation of the form $V_c=(1+\gamma)\hat{V}\cos\Omega_0 t$ and $V_s=(1-\gamma)\hat{V}\sin\Omega_0 t$ yields an image sideband of amplitude $|U_{-1}| \approx \gamma$ from the first integral in (5). Similarly, an amplitude $|U_{-1}| \approx \phi$ is produced when the phases of V_c and V_s deviate by 2ϕ from their nominal 90° phase difference. We recognize that for an image sideband suppression of, e.g. 40dB it will be necessary to make the amplitudes of V_c and V_s equal to within $\approx 10^{-2}$, and to control simultaneously their phase difference within a tolerance of $\approx 10^{-2}$rad.

In conclusion it can be stated, that the description of the frequency translator by the presented perturbational treatment provides a clearer insight into the operation of the device than our earlier matrix treatment [1], permitting direct estimates and interpretation of all critical parameters of the device.

References

1. F. Heismann and R. Ulrich, IEEE J. Quant. Electron.QE-18, 767 (1982).
2. F. Heismann and R. Ulrich, Appl. Phys. Lett. 45, 492 (1984).
3. R.C. Alferness and L.L. Buhl, Opt. Lett. 5, 473 (1980); 7, 500 (1982).

LiNbO$_3$ Optical Frequency Translators for Coherent Optical Fibre Systems

W.A. Stallard, B.E. Daymond-John, and R.C. Booth

British Telecom Research Laboratories, Martlesham Heath
Ipswich, Suffolk, United Kingdom

Mach-Zehnder interferometric frequency translators operating continuously at 1.5 µm are reported for the first time and their use in coherent optical fibre systems considered.

1   INTRODUCTION

Tunable optical frequency translators and controllers are important components in many coherent optical fibre telecommunication and sensor systems [1]. Whilst acousto-optic frequency translators can be employed, the large acoustic power requirements and the limited degree of frequency translation that can be achieved make the use of these devices in practical systems unrealistic. Recently, however, there has been considerable interest in LiNbO$_3$ Mach-Zehnder waveguide frequency translators, since frequency translations in excess of many tens of GHz are possible, the electrical power requirements can be relatively low and there is the possibility of integrating the device with other electro-optic signal processing components on a single LiNbO$_3$ wafer [2,3].

In this paper the design of Mach-Zehnder interferometric frequency translators suitable for use in coherent optical fibre systems will initially be reviewed. Experimental, results on both double and single sideband interferometric frequency translators operating continuously for the first time at a wavelength of 1.5 µm will then be reported, and finally the integration of these components with the other signal processing devices required in a coherent optical receiver will be discussed.

2   PRINCIPLE OF OPERATION OF MACH-ZEHNDER INTERFEROMETRIC FREQUENCY TRANSLATORS

If the input Y junction of a Mach-Zehnder interferometric modulator, Fig 1a, splits the incident optical power in the ratio $\alpha:(1-\alpha)$, the induced RF phase-shift in each arm is ϕ_m and there is a DC phase difference of ϕ_0 between the two modes when they recombine at the output Y junction, then it can be shown that spectral components are generated at frequencies $\omega_0 \pm n\omega_m$ and have intensities proportional to [2,3]

$$\omega_0 \pm n\omega_m: \frac{1}{2}[J_n^2(\phi_m)][1+(-1)^n 2\sqrt{\alpha(1-\alpha)}\cos(\phi_0)] \tag{1}$$

If $\alpha=0.5$ and $\phi_0=\pi$ then the fundamental and all even harmonics are suppressed with the intensity of the odd harmonics being proportional to

$$\omega_0 \pm n\omega_m: J_n^2(\phi_m) \tag{2}$$

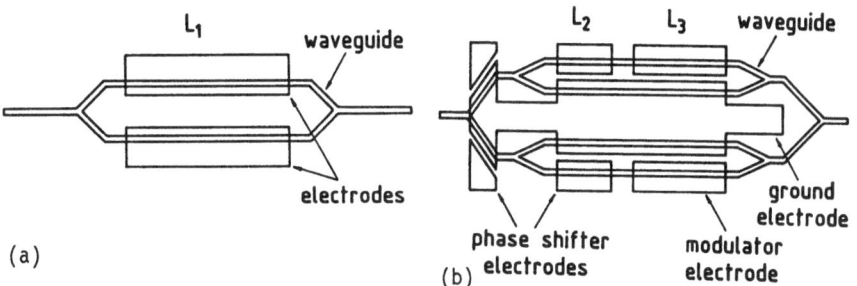

Fig.1. Mach-Zehnder frequency translators. (a) Single (L_1 = 12 mm); (b) double (L_2 = 3 mm L_3 = 8 mm)

In practice,for most depths of modulation only two sets of sidebands are significant ie $\omega_0 + \omega_m$ and $\omega_0 \pm 3\omega_m$. If ϕ_m is arranged to be 1.8 rads then the conversion efficiency of the first harmonic is maximised at -4.7 dB with that of the third harmonic being less than -20 dB. If increased frequency translation is required,then the depth of modulation can be increased to 4.2 rads so that the conversion efficiency of the third harmonic is maximised at -7.2 dB with the conversion efficiency of the first harmonic being -17.2 dB. If reduced harmonic distortion is required,the modulation depth may be reduced to ~ 3.85 rads. Whilst the conversion efficiency of the third harmonic under such conditions is then slightly reduced ie -7.6 dB the conversion efficiency of the first harmonic can in principle be zero.

If a single rather than double sideband device is required,then a double Mach-Zehnder interferometric modulator similar to that reported by IZUTSU et al [4], Fig 1b can be employed. In this device,the outputs from two appropriately biased and modulated interferometers are added,with the result that only an upper or lower sideband is generated.

3   FABRICATION

The monomode waveguide Mach-Zehnder interferometer devices shown in Fig 1 were fabricated on Z-cut LiNbO$_3$ by titanium indiffusion in a water rich atmosphere at 1050°C for 7.25 hours. The Y junction splitting half angle was 0.5° and the titanium stripe width and thickness were 7.5 μm and 73 nm respectively. Prior to the polishing of the end faces, to facilitate end fire coupling, a 200 nm thick SiO$_2$ buffer layer was deposited in order to reduce the propagation loss of the TM mode. The electrodes were fabricated from a 0.5 μm thick aluminium film and the completed device was mounted in a modified dual in-line package.

4   EXPERIMENTAL RESULTS AND DISCUSSION

4.1   SINGLE MACH-ZEHNDER FREQUENCY TRANSLATOR

The performance of the single Mach-Zehnder device, shown in Fig 1a, was initially evaluated at 400 MHz. Under DC operating conditions,the half wave voltage V_π was measured to be 7.7V and undesirable drift was not observed even after several hours of operation. The frequency-shifted output spectra were detected using a scanning etalon with a free spectral range of 7.5 GHz.

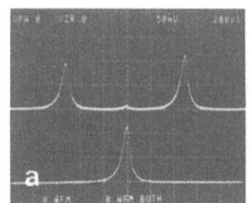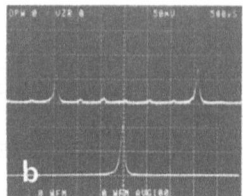

Fig.2. Single Mach-Zehnder
frequency spectra.
(lower trace) P_{RF} = 0 (50mV/div)
(upper trace) P_{RF} = 145 mw
(20 mV/div)

The spectrum obtained when modulating at 400 MHz with a DC phase bias,
ϕ_0, of π and with the depth of phase modulation adjusted to be 1.8 rads,
so as to maximise the conversion of the first harmonic, is shown in Fig
2a. It may be observed that two frequency components each translated
400 MHz from the carrier were generated. The conversion efficiency was
-5 dB, which compares favourably with a theoretical value of -4.7 dB,
and was achieved with a total RF power requirement of 145 mW.

For larger frequency translations,it is necessary to increase the
modulation depth to 3.5 rads,so that the third harmonic is generated
efficiently. A typical spectrum measured under these conditions is shown
in Fig 2b where two frequency components each translated by 1200 MHz from
the carrier may be seen. The observed maximum conversion efficiency was
-8.6 dB, compared with a theoretical value of -7.6 dB, with 600 mW of RF
drive power being required.

The experimentally measured and theoretically calculated conversion
efficiency of the frequency translated first harmonic component at 400 MHz
and third harmonic at 1200 MHz as a function of modulation depth is shown
in Fig 3. In view of the good agreement between theory and experiment,it
may be inferred that the Y junctions yield accurate 3 dB splitting with
low loss.

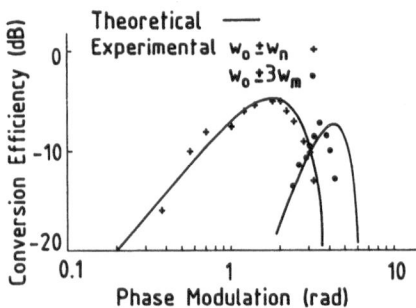

Fig.3. Frequency translator performance

4.2 DOUBLE MACH-ZEHNDER SINGLE SIDEBAND FREQUENCY TRANSLATOR

The performance of double Mach-Zehnder interferometer devices, Fig 1b, were
evaluated in a similar fashion to that described in section 4.1. For single
sideband frequency translation with suppressed carrier, the two sets of
electrodes must be driven in quadrature. For the devices described here
the close proximity of these electrodes ensured that there was sizeable
coupling between them at frequencies >100 MHz. Since it was not possible
to eliminate this coupling,it was used to provide the RF drive for one set
of electrodes whilst modulating the other set.

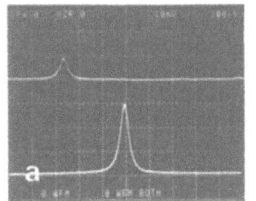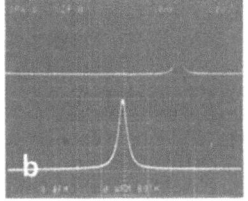

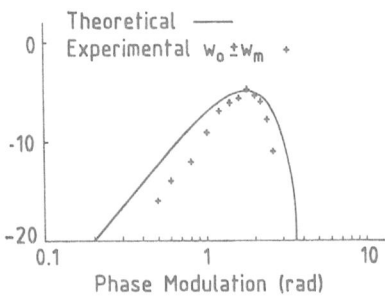

Fig.4. Double Mach-Zehnder frequency spectra.
(lower trace) fundamental
(upper trace) single sideband

Fig.5. Frequency translator performance

Single upper or lower sideband generation, Fig 4, could be achieved simply by adjustment of the DC bias voltages on the phase-shifter electrodes,and operating in this fashion the device was found to be stable for periods of many hours. The theoretically calculated and experimentally measured variation of conversion efficiency with modulation depth show remarkable agreement, Fig 5, with the maximum conversion of -4.8 dB comparing favourably with the theoretical value of -4.7 dB. For maximum conversion 540 mW or RF electrical power was required,and hence the half wave voltage V_π may be estimated as 12.8 V. This compares favourably with a value of 10 V directly measured on an adjacent test structure.

5 LiNbO$_3$ INTEGRATED OPTIC DEVICES FOR COHERENT OPTICAL RECEIVERS

Coherent optical receivers require a number of novel optical components if the full potential of these systems is to be realised. A schematic diagram of a typical coherent optical receiver is shown in Fig 6 where it may be observed that devices to perform the functions of phase, frequency and polarisation control together with optical mixing are required. Whilst each of these functions can be implemented using LiNbO$_3$ technology,if the components were packaged individually with integral fibre tails the system cost and loss would become excessive. Clearly,therefore,there are advantages to be gained from integrating all of the components on a single substrate.

One possible configuration for such a device is shown in Fig 7. The device is fabricated on Z-cut LiNbO$_3$ and a directional coupler used as an

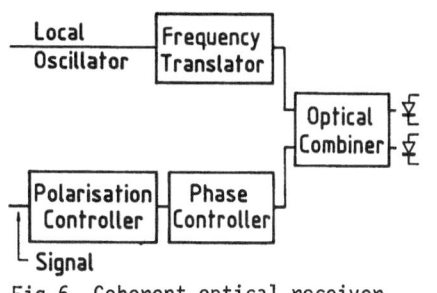

Fig.6. Coherent optical receiver

Fig.7. Integrated LiNbO$_3$ coherent receiver device

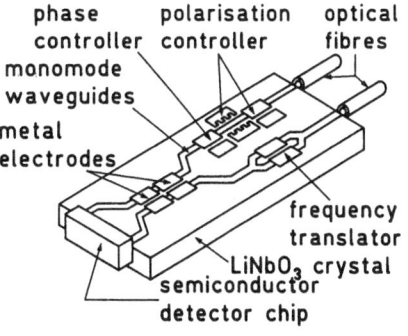

optical combiner. A polarisation controller consisting of a phase modulator and TE-TM mode convertor is used to convert the randomly polarised input signal to a linearly polarised TM mode so that it can be combined with the linearly polarised local oscillator signal in a polarisation-dependent directional coupler. (It is advantageous to use polarisation sensitive combining devices,since these devices are simpler to design and operate than devices with polarisation-independent characteristics and require considerably lower operating voltages.) The second phase modulator shown in Fig 7 is required when homodyne detection schemes are employed,and is used to match the phases of the local oscillator and incoming signal. In the device shown in Fig 7 frequency translation is achieved using a single Mach-Zehnder frequency translator in the local oscillator arm,although as has been indicated in section 2 a double Mach-Zehnder interferometer could equally well be employed if only a single frequency-shifted local oscillator signal was considered desirable.

6 CONCLUSION

The design, fabrication and performance of Mach-Zehnder interferometric frequency translators suitable for use in coherent optical fibre systems has been described. The continuous operation of both single and double Mach-Zehnder frequency translators operating at 1.5 μm has been reported for the first time. The problems of integrating frequency translating components with the other signal processing components required in a coherent receiver has been discussed,and a possible device structure proposed.

7 ACKNOWLEDGEMENTS

The authors wish to acknowledge the Director of Research, British Telecom Research Laboratories, for permission to publish this paper.

8 REFERENCES

1 R.C.Hooper, J.E.Midwinter, D.W.Smith and I.W.Stanley: J. Lightwave Tech. LT-1, 596 (1983).
2 F.Auracher and R.Keil: Appl. Phys. Lett. 36, 626 (1980).
3 C.M.Gee and G.D.Thurmond: Proc. Second European Conf. on Integrated Optics, 17-18 October 1983, Firenze Italy, pp118-121.
4 M.Izutsu, S.Shikama and T.Sueta: IEEE J. Quant. Electron. QE-17, 2225 (1981).

Low Frequency Collinear Acoustooptic TM_0-TE_0 Mode Conversion and Single Sideband Modulation in Proton-Exchanged $LiNbO_3$ Optical Waveguides

V. Hinkov, E. Ise, and W. Sohler

Universität-GH-Paderborn, Fachbereich Physik, Angewandte Physik, Postfach 1621 D-4790 Paderborn, Fed. Rep. of Germany

It is shown that the birefringence of Ti-diffused/proton exchanged optical waveguides in $LiNbO_3$ can be precisely controlled to adjust phase matching for collinear acoustooptic mode conversion, also for low acoustic frequencies. Fabrication and properties of a 85 MHz mode converter/single sideband modulator are reported.

1) Introduction

There is a considerable interest in integrated acoustooptical devices for signal processing applications [1]. Among them, devices with collinear acoustooptic interaction are especially attractive, due to their simple structure. In particular, collinear acoustooptic mode conversion has been investigated theoretically as well as experimentally [2, 3]. The frequency of the surface acoustic wave (SAW) used for such an interaction is determined by the phase match condition for the interacting waves, and therefore by the natural birefringence of the ($LiNbO_3$) substrate (which is only slightly modified by the Ti-indiffusion for the waveguide fabrication). For this reason, the center frequency of $LiNbO_3$ devices is about 500 MHz if operated with light of a wavelength $\lambda = 0.63$ μm.

Recently, proton exchange of Li-ions in $LiNbO_3$ was reported as a new method to fabricate high index waveguides and to modify in this way the birefringence of the material [4, 5]. Following this work, we could control the birefringence of optical waveguides in $LiNbO_3$ precisely by a combination of Ti-indiffusion and proton exchange with a subsequent annealing procedure (section 2). In this way, low acoustic frequency applications become possible. We could demonstrate for the first time collinear acoustooptic mode conversion with a SAW of low (compared to 500 MHz) frequency (section 3). Furthermore, the device properties were investigated when used as a single sideband modulator (section 4).

2) Sample Preparation

All our experiments were performed with Y-cut $LiNbO_3$ samples, the optical guided waves and the surface acoustic waves (SAW) propagating collinearly in X-direction; this is the most favourable interaction geometry. To prepare a sample, a 250 Å thick Ti-film was indiffused first at $1060°C$ for 8 hours in an oxygen atmosphere. In this (conventional) way an optical waveguide was fabricated, supporting two modes of each polarization at the wavelength $\lambda = 0.63$ μm. Then an exchange of Li-ions by protons followed: it was performed in pure benzoic acid for 18 hours at a temperature of $250°C$ using a vacuum-sealed quartz ampoule to avoid contamination [6]. By this process, only the extraordinary index of refraction is considerably increased inside the exchanged surface layer, thereby producing an optical multimode guide for TE-polarization; the ordinary index of refraction is slightly lowered. The diffusion profiles of the superimposed Ti-diffused/proton exchanged optical waveguides with losses of about 1.5 dB/cm were reconstructed from their mode spectra using the inverse WKB-method [7]. The n_e-profiles were step-like (the WKB-method cannot resolve the flat "Ti-profile" ($\Delta n \sim 0.005$) superimposed on the high step ($\Delta n \sim 0.1$) produced by the proton exchange.) The n_o-profiles could not be resolved; only the effective indices of the (2 or 1) guided modes were measured (see Fig. 1).

Following the procedure described in [8], the samples were then annealed for some hours in an oxygen atmosphere at 400°C to modify the birefringence of the fundamental modes. After each annealing process the index profiles respectively the mode indices were measured again (see Fig. 1). The control of the annealing time allowed to adjust the birefringence precisely. Fig. 2 presents the difference $\Delta n_{eff} = n_{eff}(TE_o) - n_{eff}(TM_o)$ of the effective indices of the fundamental TE and TM modes as function of the annealing

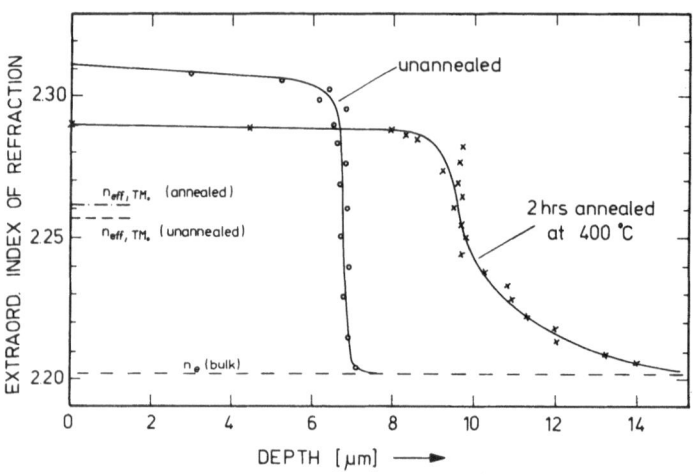

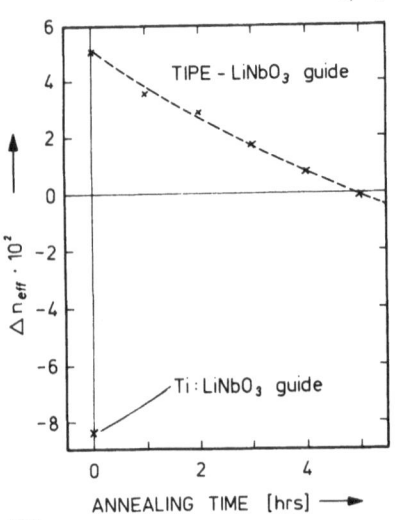

Fig. 1: Profiles of the extraordinary index of refraction of a Ti-diffused/ proton exchanged optical waveguide before and after 2 hours of annealing at 400°C reconstructed via the WKB-method

Fig. 2: Birefringence of the fundamental modes of a Ti-diffused/proton exchanged (TIPE) optical waveguide and of a Ti:LiNbO$_3$ guide as function of the annealing time ($\lambda = 0.63$ μm)

170

time. Without annealing, the birefringence of the Ti:LiNbO$_3$ guide is nearly inverted. It is remarkable that even guides of zero birefringence ("isotropic" guides) can be fabricated in this way.

3) Acoustooptic TM$_0$-TE$_0$ Mode Conversion

To achieve a TM$_0$-TE$_0$ mode conversion by a collinear interaction of an optical guided wave and a surface acoustic wave phase matching is necessary. The corresponding wave vector condition $k(TM_0) + K(SAW) = k(TE_0)$ determines the wavelength of the SAW yielding $1/\Lambda = \Delta n_{eff}/\lambda$ (Λ is the wavelength of the SAW). With the possibility to adjust Δn_{eff} as described in the preceding section, any acoustic wavelength larger than approximately 13 μm can be chosen to achieve phase-matched mode conversion. This corresponds to a SAW frequency between 0 and about 260 MHz; of course the interaction efficiency depends on the overlap of optical and acoustic wave, and therefore decreases with increasing Λ.

To demonstrate the possibility for a drastic reduction of the SAW frequency, we performed experiments with 85 MHz (remember that a collinear interaction in Ti:LiNbO$_3$ guides requires a frequency of about 500 MHz at $\lambda = 0.63$ μm). The corresponding acoustic wavelength was 40 μm necessitating a birefringence $\Delta n_{eff} = 0.016$. Such samples were fabricated by an annealing procedure of 3 hours. The SAW was excited and detected by means of interdigital transducers consisting of 4.5 fingerpairs with finger-width and -spacing of 10 μm. The aperture of the acoustic beam was 3 mm. The transducer used for SAW excitation was formed on an unexchanged part of the waveguide as the proton exchange reduces the excitation efficiency [9]. The acoustic properties of the Ti-diffused/proton exchanged samples could be compared with a Ti-diffused region beside fabricated on the same substrates. A diagram of such a sample is shown in Fig. 3.

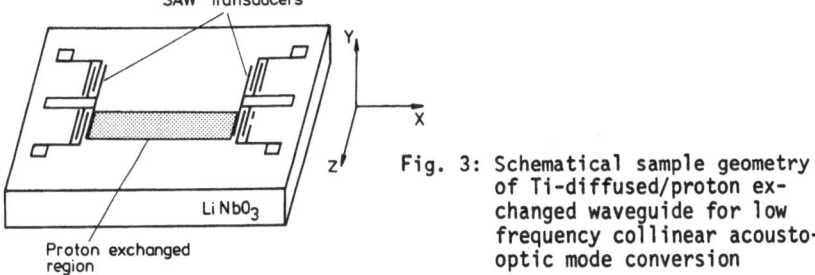

Fig. 3: Schematical sample geometry of Ti-diffused/proton exchanged waveguide for low frequency collinear acoustooptic mode conversion

Via rutile prisms the light beam of a He-Ne laser was coupled into and out of the waveguide. We excited the fundamental TM mode; along an interaction length of 15 mm acoustooptic mode conversion to the TE$_0$ mode occurred. The maximum conversion efficiency was 5 % with 4 W input electrical power. It is of course possible to improve the efficiency several times, e.g. by optimizing the transducer geometry, by increasing the interaction length or by fabricating an acoustic channel guide.

4) Single Sideband Modulation

The acoustooptical interaction is not only determined by wavevector (quasi-momentum) conservation but also by energy conservation. As a consequence, the frequency of the TE$_0$ mode (converted from the input TM$_0$ mode) is shifted by the acoustic frequency Ω according to $\hbar\omega(TM_0) + \hbar\Omega = \hbar\omega(TE_0)$. The acousto-optic mode converter can therefore be used as a frequency shifter or a single sideband modulator. Such a device is of great interest for several signal processing applications; in the form of a bulk acoustooptical Bragg modulator it was used e.g. for signal detection in a fiber-optic gyroscope.

To investigate the spectral shift respectively the carrier frequency suppression, the decoupled light of the converted TE_0 mode was analyzed by a scanning Fabry-Perot interferometer with a free spectral range of 955 MHz. The results (with low and high resolution) are given in Fig. 4.

The small peak arises from a TE_0 mode of unshifted frequency (the optical carrier frequency) produced by imperfections of the waveguide; it gives the zero of the frequency scale. The large peak is due to the frequency shifted TE_0 mode generated by the acoustooptic interaction. According to theory, a frequency shift of 85 MHz was observed. The ratio of the intensities of the sideband and the carrier was 5 : 1 to 10 : 1 depending on the lateral beam position in the planar waveguide; it is evident that this ratio can be improved too by an enhancement of the conversion efficiency.

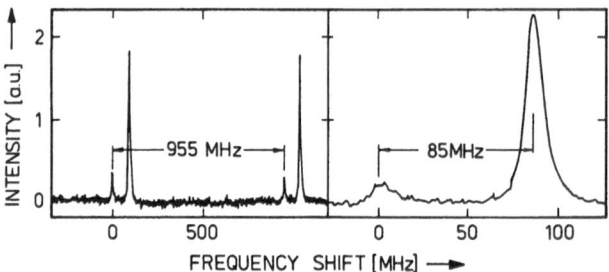

Fig. 4: Scanning Fabry-Perot spectra of the output of the single sideband modulator; left: low resolution; right: high resolution. Free spectral range is 955 MHz

5) Conclusion

In conclusion, a low-frequency acoustooptic TM-TE mode converter or single sideband modulator was fabricated. A very simple device structure with collinear acoustooptic interaction could be chosen by tailoring a Ti-diffused/proton exchanged ("TIPE") waveguide to achieve phase matching. The decrease of the SAW velocity on the proton exchanged substrate [9] will allow to construct doubly confined structures (channel guides) for the acoustical and optical wave. As a result, the mode conversion efficiency will be increased as well as the frequency bandwidth of the acoustooptic interaction.

Acknowledgment

V. Hinkov would like to thank the Alexander von Humboldt foundation for the financial support.

References

1 C.S. Tsai: Proc. 1st Europ Conf. Integrated Optics, London, 1981, IEE conf. publication no 201, p. 87

2 K. Yamanouchi,K. Higuchi and K. Shibayama: Appl. Phys. Lett. $\underline{28}$, 75 (1976)

3 L.N. Binh and J. Livingstone: IEEE J. Quant. Electron. QE-$\underline{16}$, 964 (1980)

4 J.L. Jackel, C.E. Rice and J.J. Veselka: Appl. Phys. Lett. $\underline{41}$, 607 (1982)

5 J. Botineau, S. Neveu, P. Sibillot, D.B. Ostrowsky and M. De Micheli:
   Opt. Lett. $\underline{8}$, 114 (1983)

6 M. De Micheli, J. Vollmer, J.P. Baretti and S. Neveu: Proc. 2nd Europ.
   Conf. Integrated Optics; post deadline paper; Firenze 1983; IEE conf.
   publication no 227

7 J.M. White and P.F. Heidrich: Appl. Opt. $\underline{15}$, 151 (1976)

8 V. Hinkov and E. Ise: to be published

9 V. Hinkov and E. Ise: to be published

A Simple and Wide Optical Bandwidth TE/TM Converter Using Z Propagating LiNbO$_3$ Waveguides

C. Mariller and M. Papuchon

THOMSON-CSF Laboratoire Central de Recherches, Domaine de Corbeville, B.P. 10 F-91401 Orsay Cedex, France

Efficient TE/TM electrooptic conversion in Z propagating LiNbO$_3$ waveguides is reported. 95 % conversion is obtained for a command voltage of 20 Volts. The interaction does not require periodic electrodes and thus leads to large bandwith devices.

SUMMARY

LiNbO$_3$ is a particularly interesting substrate to realize active electrooptic devices especially using the Ti indiffusion technique. Among the various devices which have been reported the TE/TM converter [1] can lead to very interesting applications such as polarization controllers [2], polarization independant devices [3] or frequency shifters [4,5]. In addition, the association of such a converter with an integrated polarizer will give the possibility of realizing efficient amplitude modulators. In general the substrate orientation used in the above experiments leads to very high birefringent waveguides and in order to obtain an efficient interaction, periodic coupling is needed. The TE/TM coupling is achieved via a non diagonal electrooptic coefficient which is usually r_{51}. In this case, it is necessary to use periodic electrodes to realize the phase matching condition. As the birefringence of LiNbO$_3$ is of the order of 0,1 the period needed is around 10 μm leading, for an interaction length of the order of 1 cm, to very narrow bandwith devices.

In this paper we describe the operation of a TE/TM converter using a non standard substrate orientation (Z propagating waveguides) which permits to obtain very small TE/TM birefringence and thus to use a very simple non periodic electrode to apply the electric field. As a consequence, the optical bandwith is very large and in addition the optical damage and the out diffusion problems are drastically reduced. The main drawback of the new configuration is the use of an electrooptic coefficient (r_{61}) smaller than usual.

The waveguide configuration is shown in Fig. 1. The waveguides lie along the Z axis of the crystal and the Y axis is perpendicular to the substrate. When an horizontal electric field is used the TE $\rightleftarrows$ TM conversion is achieved via the r_{61} electrooptic coefficient. Using the configuration described above, the TE/TM birefringence can be very small and a periodic electrode system is not needed (the TE and TM modes are quasi degenerated). In our experiment a 10 mm long electrode system is used with a spacing of around 7 μm. The single mode waveguides are realized by the classical Ti : indiffusion technique (Ti strip width : 3,2 μm, Ti thickness : 500 A) at a temperature of 1050°C for 8 hours. The edges are then polished to permit direct butt coupling and polarized light coming from a semiconductor laser diode emitting in the 0,8 μm region is used a source.

The experimental results are shown on the Fig. 2 and 3, where the output intensities in the TE and TM polarizations are shown versus the applied voltage.

174

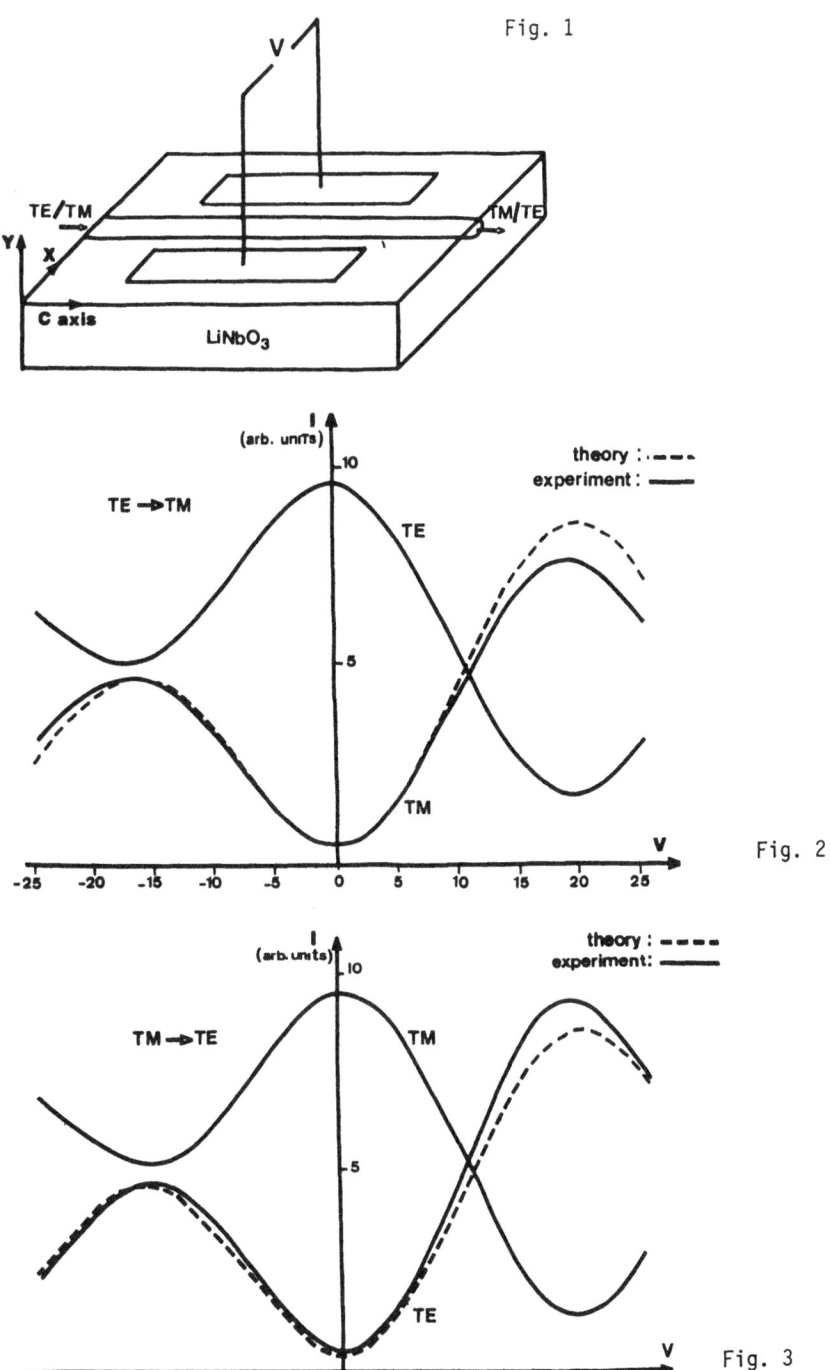

Fig. 1

Fig. 2

Fig. 3

Figures 2 and 3 correspond to inputs in the TE and TM polarization respectively.

As can be seen in these figures the response is not the same for positive or negative values of the applied voltages. This can be explained by the residual difference between the two propagation constants of the TE and TM modes in addition with a voltage induced birefringence probably due to a slight misalignment of the electrodes. More than 90 % conversion efficiency is observed at an applied voltage of + 18 Volts. In these experiments the source was a semiconductor laser emitting at $\lambda = 0,84$ µm. To qualitatively measure the optical bandwith, a superluminescent diode was then coupled into the waveguide. On the bandwith of the source ($\Delta\lambda \simeq 270$ A) no noticeable change in the conversion efficiency/command voltage was observed indicating the large bandwith capability of the device. To estimate the coupling constants, and the induced and static TE/TM birefringence the classical loss less coupled mode theory was used. A typical theoretical fitting curve is shown in Fig. 2 and 3 for which the parameters were taken as follows :

coupling constant K $= 7,2$ m^{-1} Volt^{-1}

Initial ($\Delta\beta_{TE/TM/_{Ko}}$) $= 1,3 \ 10^{-5}$

Induced ($\Delta\beta_{TE/TM/_{Ko}}$) $= 1,27 \ 10^{-6}$ Volt^{-1}

In conclusion, we have demonstrated a TE/TM converter using a non standard orientation for the LiNbO$_3$ substrate. The device has several interesting properties like the use of a very simple non periodic interaction, large optical bandwith ($>$ 270 A) and a good conversion efficiency ($>$ 90 %) for an applied voltage of around 20 Volts. The Z propagating waveguides used here are obviously free from classical out diffusion problems and must be less sensitive to optical damage. The main drawback is the use of a relatively small electrooptic coefficient (r_{61}). This can be partly overcome by optimizing the electrode configuration. In addition it must be noted that this configuration can lead to extremely simple amplitude modulators by using integrated polarizers and can be implemented in many devices where TE $\rightleftarrows$ TM conversion is the basic effect.

Acknowledgements

The authors acknowledge J.M. Arnoux, D. Papillon and M. Werner for their expert technical assistance, S. Vatoux, Y. Bourbin and K. Thyagarajan for fruitful discussions and the Laboratoire Semiconducteurs pour l'Optoélectronique for the realization of the semiconductor sources used in their experiments.

REFERENCES

1 R.C. Alferness and L.L. Buhl, Opt. Lett., 5, 473 (1980).
2 R.C. Alferness and L.L. Buhl, Appl. Phys. Lett., 38, 655 (1981).
3 R.C. Alferness and L.L. Buhl, Appl. Phys. Lett., 39, 131 (1981).
4 L.M. Johnson, R.A. Becker and R.H. Kingston, paper WD4, 7th Topical Meeting on Integrated and Guided wave Optics Kissimmee (April 1984).
5 F. Heismann and R. Ulrich, Appl. Phys. Lett., 45, 490 (1984).

Part V

Fundamentals and Wave Guiding

Nonlinear Third Order Integrated Optics

Colin T. Seaton, and George I. Stegeman

Optical Sciences Center and Arizona Research Laboratories,
Tucson, AZ 85721, USA

W.M. Hetherington III

Department of Chemistry, University of Arizona, Tucson, AZ 85721, USA

H.G. Winful

GTE Laboratories Incorporated, 40 Sylvan Road, Waltham, MA 02254, USA

1. Introduction

The nonlinear mixing of optical beams has led to a rich spectrum of phenomena since the inception of the field in the early 1960's. This includes the generation of new frequencies, a host of nonlinear spectroscopies, and a variety of signal processing operations including phase conjugation, optical bistability, and optical switching. In general, nonlinear optical interactions occur whenever the optical fields associated with one or more laser beams propagating in a material are large enough to produce polarization fields proportional to the product of two or more of the incident fields [1]. These nonlinear polarization fields radiate fields at the nonlinear frequency that grow linearly with propagation distance under optimum conditions of phase-matching. Hence the key to obtaining efficient nonlinear optical interactions is to maintain high optical intensities over as long a distance as possible.

Many characteristics of optical waveguides suggest that they should be optimum media for performing efficient nonlinear interactions. The strong beam confinement to regions of the order of the wavelength of light in one (planar waveguides) or two (channel waveguides) dimensions implies that large intensities can be produced with small total powers. Diffractionless propagation (in the confined dimension) occurs down the film for centimeter distances, limited by absorption and/or scattering. Furthermore, phase-matching in this format can be implemented in the plane of the surface. Since nonlinear interactions can take place in either the film or the bounding media, waveguide dispersion can be used to achieve phase-matching in materials that are not phase-matchable with plane wave fields. Therefore, efficient all-optical devices can potentially be implemented with semiconductor lasers, an improvement over the use of plane wave fields that require high power lasers.

These features were recognized in the early days of integrated optics and harmonic generation was investigated by a number of groups [2-5]. The control of LiNbO₃ waveguides has now led to the demonstration of efficient operations such as harmonic generation [6,7], parametric mixing [8], parametric oscillators [9], and signal processing [10,11], many of which are discussed in detail elsewhere in these proceedings. All of these interactions depend on $\chi^{(2)}$, the second order susceptibility.

New developments in nonlinear optics in the last few years have centered on phenomena depending on $\chi^{(3)}$, the third order susceptibility. The starting point for any nonlinear interaction is the expansion of the nonlinearly produced polarization caused by the presence of multiple fields [1]. The nonlinear polarization is usually written as

$$P^{NL} = \epsilon_0 \chi^{(3)} : EEE + .. \tag{1}$$

178

where $\chi^{(3)}$ is third order susceptibility, and E is the total field. For frequency inputs at ω_a, ω_b, and ω_c (some of which may be degenerate), the fields are written, for example, as

$$E^a = \frac{1}{2}E^a(z)e^{i(\omega t - \beta k_0 x)} + c.c. \tag{2}$$

for guided waves propagating along the x-axis with z normal to the surfaces, and $k_0 = \omega/c$. Therefore the nonlinear polarization, and hence the radiated fields, can have the frequency components $\omega_a \pm \omega_b \pm \omega_c$, which leads to a large range of phenomena. In contrast to efficient second order phenomena that are difficult to phase-match, there are third order processes such as degenerate four-wave mixing [12] and optical bistability [13] that are automatically phase-matched. There are interactions in which tuning the difference frequency between two of the laser fields through characteristic molecular vibrational frequencies leads to resonant enhancements in the signal and hence can be used for spectroscopy [14]. Here we discuss guided wave versions of these phenomena. In addition there are third order interactions that are unique to integrated optics, and have no plane wave analogs. In particular, there is the power-dependent coupling into waveguides via distributed couplers such as prism and gratings, and a new class of guided waves whose properties change dramatically with guided wave power. In this paper we discuss nonlinear phenomena in waveguides that utilize the third order susceptibility, concentrating on work carried out in our laboratory.

2. Power Dependent Refractive Index

The simplest case is that of only one beam present at frequency $\omega(=\omega_a)$ in an isotropic medium. The polarization term which oscillates at ω is given by [15]

$$P_{\gamma i}^{NL}(z) = c\epsilon_0^2 n_\gamma^2 n_{2\gamma}\left[\frac{2}{3}E_{\gamma i}(z)E_{\gamma j}(z)E_{\gamma j}^*(z) + \frac{1}{3}E_{\gamma i}^*(z)E_{\gamma j}(z)E_{\gamma j}(z)\right] \tag{3}$$

where $n = n_\gamma + n_{2\gamma}S$ for the γ'th medium, S is the local intensity, and $n_{2\gamma}$ is the intensity-dependent refractive index. Note that we reserve the subscript $\gamma = c$, f or s for identifying the medium as either the cladding, film or substrate respectively. For TE polarized waves [16]

$$P_{\gamma y}^{NL}(z) = c\epsilon_0^2 n_\gamma^2 n_{2\gamma}|E_{\gamma y}(z)|^2 E_{\gamma y}(z) \tag{4a}$$

The situation for TM is much more complicated, because there are two field components, E_x and E_z, that have a non-zero phase difference between them. The simplest case occurs for either the substrate or the cladding medium where $E_x(z)$ and $E_z(z)$ are $\pi/2$ out of phase with one another; evaluating (3) [16],

$$P_{\gamma x}^{NL} = c\epsilon_0^2 n_\gamma^2 n_{2\gamma}\left[|E_{\gamma x}(z)|^2 + \frac{1}{3}|E_{\gamma z}(z)|^2\right]E_{\gamma x}(z), \tag{4a}$$

$$P_{\gamma z}^{NL} = c\epsilon_0^2 n_\gamma^2 n_{2\gamma}\left[|E_{\gamma z}(z)|^2 + \frac{1}{3}|E_{\gamma x}(z)|^2\right]E_{\gamma z}(z). \tag{4b}$$

Writing the dielectric displacement field as,

$$D_{\gamma i}(z) = \epsilon_0 n_\gamma^2 E_{\gamma i}(z) + \epsilon_0 \alpha_{ij}|E_{\gamma j}(z)|^2 E_{\gamma i}(z), \tag{5}$$

and restricting the discussion to evanescent fields with either a TE or a TM wave present

$$\alpha_{xx} = \alpha_{zz} = \alpha_{yy} = c\epsilon_0 n_\gamma^2 n_{2\gamma}, \tag{6a}$$

$$\alpha_{xz} = \alpha_{zx} = \frac{1}{3}c\epsilon_0 n_\gamma^2 n_{2\gamma}, \tag{6b}$$

Clearly one of the consequences of the third order nonlinearity is a field dependent refractive index, and therefore an intensity-dependent guided wave wavevector.

3. Nonlinear Guided Waves

The effect of an intensity-dependent refractive index on the propagation properties of a guided wave can either be calculated approximately using coupled mode theory [16], which assumes that the field distribution is not affected by the nonlinearity, or exactly by solving the nonlinear wave equation and boundary conditions [17-23]. Coupled mode theory approach can be used when the maximum change in index that can be produced optically, $\Delta n_{sat} \geq n_{2,E}|E|^2$, is less than the index differences $n_f - n_c$ and $n_f - n_s$ that exist at low powers between the film and the bounding media. Otherwise, the field distributions are also affected and the more exact theory must be used.

A great deal of progress has been made recently in solving the nonlinear wave equation [24] as applied to thin film waveguides [18-23]. The solutions for the TE case are exact; the appropriate formulation for the TM case is still in dispute [25]. Assuming a TE wave of the form given by (2), the nonlinear wave equation for $E_{\gamma y}(z)$ is

$$\nabla^2 E_{\gamma y}(z) + k_0^2[n_\gamma^2 + \alpha_\gamma E_{\gamma y}(z)^2]E_{\gamma y}(z) = 0. \tag{7}$$

The solutions to this equation are by now well-known [24]. In a nonlinear cladding

$$E_{cy}(z) = \sqrt{\frac{2}{\alpha_c}} \frac{q}{\cosh[k_0 q(z_1-z)]} \tag{8a}$$

$$E_{cy}(z) = \sqrt{\frac{2}{|\alpha_c|}} \frac{q}{\sinh[k_0 q(z_1-z)]} \tag{8b}$$

for $n_{2c} > 0$ (self-focusing medium) and $n_{2c} < 0$ (self-defocusing medium) respectively with $\alpha_\gamma = n_\gamma^2 c\epsilon_0 n_{2\gamma}$. Here $q^2 = \beta^2 - n_c^2$ and z_1 is a constant that can be calculated from the total guided wave power per unit distance along the wavefront (discussed later).

In the general case, both the film and substrate can also be nonlinear. Here we concentrate on the case that we have studied experimentally [26], namely film and substrate media with intensity-independent refractive indices and $n_{2c} > 0$. The fields inside the film are written in the usual way as a superposition of sine and cosine functions ($n_f > \beta$) or sinh and cosh functions ($\beta > n_f$, now allowed for some cases) with argument $k_0\kappa z$ where $\kappa^2 = |\beta^2 - n_f^2|$. When the continuity of tangential electric and magnetic fields is applied

$$E_{fy}(z) = E_{cy}(0)\left[\cos(k_0\kappa z) + \frac{q}{\kappa}\tanh(k_0 q z_1)\sin(k_0\kappa z)\right] \tag{9a}$$

$$E_{fy}(z) = E_{cy}(0)\left[\cosh(k_0\kappa z) + \frac{q}{\kappa}\tanh(k_0 q z_1)\sinh(k_0\kappa z)\right] \tag{9b}$$

for $\beta^2 < n_f^2$ and $\beta^2 > n_f^2$ respectively. The substrate fields are written in the usual way as $E_{ys}(z) = E_{yf}(h)\exp[-sk_0(z-h)]$ that leads to the dispersio relation

$$\tan(k_0\kappa h) = \frac{\kappa[q\tanh(k_0qz_1) + s]}{\kappa^2 - sq\tanh(k_0qz_1)} \tag{10a}$$

$$\tanh(k_0\kappa h) = \frac{\kappa[q\tanh(k_0qz_1) + s]}{-\kappa^2 - sq\tanh(k_0qz_1)} \tag{10b}$$

for $\beta^2 < n_f^2$ and $\beta^2 > n_f^2$ respectively. Note that the limit $n_{2c} \to 0$ corresponds to $z_1 \to \infty$, in which limit one recovers the usual dispersion relationships.

What remains is to determine the constant z_1 from the guided wave power, or vice-versa. The guided wave power per unit length along the y-axis is obtained in the usual way by integrating the Poynting vector over the depth dimension, i.e.,

$$P = \int_{-\infty}^{\infty} \mathbf{ExH}dz = P_c + P_f + P_s .$$

For the cladding this gives

$$P_c = \frac{\beta q}{k_0 n_c^2 n_{2c}} [1 - \tanh(k_0qz_1)] , \qquad n_{2c} > 0 \tag{11}$$

for TE modes and all values of β. For the film,

$$P_f = \frac{\beta q^2}{2n_c^2 n_{2c}} \frac{1}{\cosh^2(k_0qz_1)} [h(1+\frac{q^2}{\kappa^2}\tanh^2(k_0qz_1))$$

$$+ \frac{\sin(2k_0\kappa h)}{2k_0\kappa}(1-\frac{q^2}{\kappa^2}\tanh^2(k_0qz_1)) + \frac{q}{\kappa^2 k_0}\tanh(k_0qz_1)(1-\cos(2k_0\kappa h))] \tag{12}$$

for $n_f^2 > \beta^2$, and,

$$P_f = \frac{\beta q^2}{2n_c^2 n_{2c}} \frac{1}{\cosh^2(k_0qz_1)} [h(1-\frac{q^2}{\kappa^2}\tanh^2(k_0qz_1))$$

$$+ \frac{\sinh(2k_0\kappa h)}{2k_0\kappa}(1+\frac{q^2}{\kappa^2}\tanh^2(k_0qz_1)) + \frac{q}{\kappa^2 k_0}\tanh(k_0qz_1)(\cosh(2k_0\kappa h)-1)] \tag{13}$$

for $\beta^2 > n_f^2$. For the substrate,

$$P_s = \frac{\beta q^2}{2k_0 s n_c^2 n_{2c}\cosh^2(k_0qz_1)}\left[\cos(k_0\kappa h)+\frac{q}{\kappa}\tanh(k_0qz_1)\sin(k_0\kappa h)\right]^2 \tag{14}$$

and

$$P_s = \frac{\beta q^2}{2k_0 s n_c^2 n_{2c}\cosh^2(k_0qz_1)}\left[\cosh(k_0\kappa h)+\frac{q}{\kappa}\tanh(k_0qz_1)\sinh(k_0\kappa h)\right]^2 \tag{15}$$

for $n_f^2 > \beta^2$ and $\beta^2 > n_f^2$ repsectively.

We have evaluated [22] these dispersion relations for a case similar to that studied experimentally; a glass film of $n_f = 1.57$, a cladding medium with $n_c = 1.55$ corresponding to the liquid crystal MBBA, which has

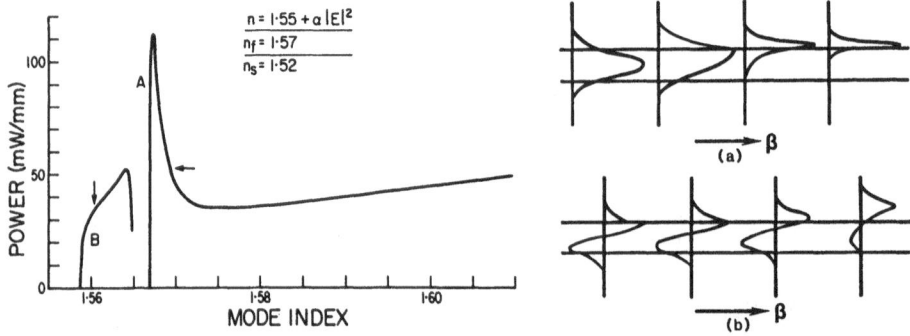

Fig. 1. The guided wave power versus effective index for nonlinear TE_0 and TE_1 waves guided by a 2.0-μm-thick film. The material parameters are $n_f=1.57$, $n_c=1.55$, $n_s=1.52$, and $n_{2c}=10^{-9}$ m²/W.

Fig. 2. The field distributions associated with the nonlinear TE_0 (a) and TE_1 (b) waves in Fig. 1.

$n_{2c}=10^{-9}$m²/W at $\lambda=0.515$ μm and a substrate with $n_s=1.52$. In Fig. 1, A and B refer to the TE_0 and TE_1 modes respectively. For small powers, the effective index varies linearly with guided wave power, and the results agree with the predictions of the approximate coupled mode approach. However, when the optically induced change in cladding index becomes comparable to n_f-n_c, the field distributions begin to change (Fig. 2), the field maximum moves into the self-focussing medium, which has dramatic effects on the effective index β. For the TE_0 mode, solutions exist for $β>n_f$ and degenerate into nonlinear surface polaritons guided by a single interface in the high power region, where β increases monotonically with power. For the TE_1 mode, one of the field extrema crosses into the cladding at the arrow, and this mode also becomes progressively more localized in the cladding. Note that there is an absolute maximum power that can be transmitted via this mode. Furthermore, in both cases there are regions where two versions of the same mode can coexist at a given power level, suggesting the possibility for bistable operation.

The experimental configuration is as follows [26]. A bead of liquid crystal MBBA is placed between two coupling prisms on the surface of a glass-on-glass thin-film waveguide. TE_0 and TE_1 modes were coupled in separate experiments into the waveguide with strontium titanate prisms, guided through the region with the liquid crystal on top, and were coupled out with a second strontium titanate prism. The results are shown in Fig. 3. In both cases, the transmission through the system is typically 1% due to coupling inefficiencies, reflections at the two liquid crystal boundaries and scattering and absorption losses. For the TE_0 case, the transmitted power is linear in the incident power with some saturation effects evident at the highest power levels investigated. On the other hand, in the TE_1 case there is a pronounced saturation effect, as well as hysteresis with respect to increasing versus decreasing incident power. The results were reproducible, and time periods in excess of 30 sec were taken after any incident intensity change to ensure equilibrium. By varying the input intensity rapidly, we determined the time constant for the process to be ≈1 sec.

Increasing the power initially moves the transmission along the upper branch. When the maximum power transmission point is reached, the

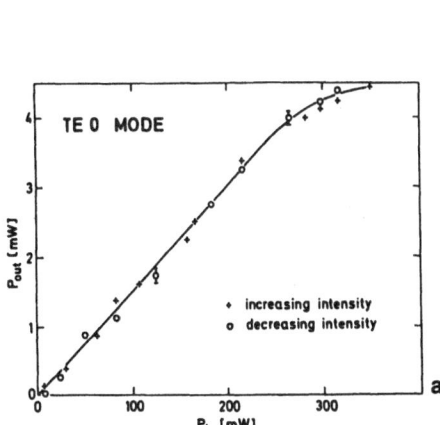
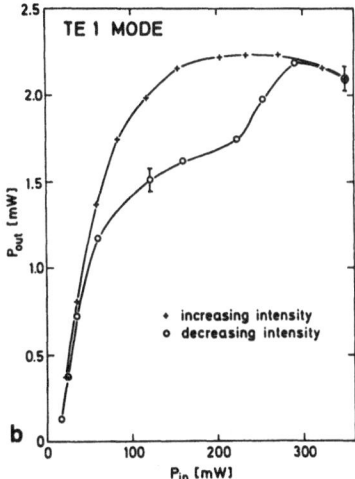

Fig. 3. Transmitted versus incident power for nonlinear TE_0 (a) and TE_1 (b) guided waves.

transmitted power saturates as indicated in Fig. 3b. Now, as the guided wave power is decreased, waves on both the upper and lower branches are excited, and hence the net transmission drops. Because the field becomes progressively more localized in the liquid crystal as power is decreased along the lower branch, progressively less of this mode is excited at the liquid crystal-air boundary as the prism-launched TE_1 wave encounters the liquid crystal bead. This results in the transmission curve approaching the linear transmission curve. We concluded that the experimental curves are described well by the theory, and that nonlinear guided waves have been excited in this experiment. Clearly this type of phenomenon has applications to broadband optical limiters.

There are many possible applications of this phenomenon, especially when both bounding media are nonlinear. These will appear in print in the near future.

4. Power-Dependent Distributed Couplers

The coupling of an external radiation field into a guided wave is usually achieved by distributed couplers such as prisms or gratings. Phase-matching is required for these devices to be efficient. In the prism case, the projection of the plane wave wavevector in the prism onto the base of the prism must equal the guided wave wavevector. However, it was shown above that the guided wave wavevector changes with guided wave power if one of the guiding media has an intensity-dependent refractive index. Therefore the phase-matching condition changes as the guided wave power grows under the prism and, for a fixed angle of incidence for the input beam, the coupling efficiency becomes power-dependent and decreases [27].

The operation of such a nonlinear coupler was described in detail in references [27,28]. We now write the field term $E_y(z)$ in (2) as

$$E_y(z) = Cf(z)a(x) \tag{16}$$

where C is a normalization constant chosen so that $|a_g(x)|^2$ is the guided wave power in watts per meter. The coupling between the guided wave and incident fields is described by [29]

$$\frac{\partial a(x)}{\partial x} = \hat{t}a_{inc}(x)e^{i\Delta\beta(|a[x]|^2)k_0 x} - (\ell^{-1}+\alpha)a(x) \quad \text{with} \tag{17}$$

$$\Delta\beta = n_p\sin\theta - \beta(|a[x]|^2), \tag{18}$$

where $\hat{t}$ is a transfer coefficient, α is the waveguide attenuation coefficient and ℓ is the characteristic distance for reradiation of the guided wave field into the prism [29]. Assuming that the guided wave power remains small enough to remain in the linear region of Fig. 1,

$$\beta(|a[x]|^2) = \beta_0 + \Delta\beta_0'|a(x)|^2. \tag{19}$$

The field incident onto the prism-waveguide interface at an angle θ to the normal has a field distribution at the base of the prism (refractive index n_p) given by $a_{inc}(x) \propto \exp(-x^2/w_0^2)$. Assuming that $\Delta\beta$ can be set to zero at low powers by adjusting the coupling conditions, then $\Delta\beta=\Delta\beta_0'|a(x)|^2$. Therefore, as the guided wave power increases, the synchronous coupling condition $\Delta\beta=0$ is no longer valid, phase-mismatch occurs between the incident and guided wave fields and the coupling efficiency is reduced. A detailed calculation of this effect was given in reference [28].

The experiment [30] to verify the power-dependent coupling efficiency was a modification of that used to demonstrate the existence of nonlinear waves. The liquid crystal MBBA filled the gap between the input strontium titanate prism and the waveguide. (In the previous case, the liquid crystal was placed on the surface between the two prisms.) When a few milliwatts of 0.515 µm light from an argon ion laser is transmitted through the sample, the results shown in Fig. 4 are obtained. The salient feature is that under certain conditions, the transmitted intensity is not linear with the input intensity.

The power levels at which nonlinear coupling should occur can be estimated from the phase-mismatch term in (18). The maximum phase-shift is

$$\Delta\phi \approx \frac{\Delta\beta_0'}{k_0}\int_{-\infty}^{\infty}|a(x)|^2dx \approx 2w_0\eta\Delta\beta_0'k_0^{-1}P_{inc}$$

where P_{inc} is the incident power per unit length in the prism and η is the optimized low power coupling efficiency. For phase-shifts of the order of

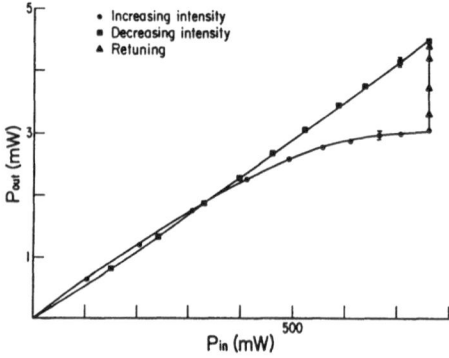

Fig. 4. Experimental results for the coupled-in versus incident power for measurements taken over time intervals long compared to the material relaxation time

π/2 or more, the coupling efficiency becomes power dependent. Note that although this phase-shift can be reduced by reducing w_0, if w_0 is made smaller than the characteristic coupling length [29] then η is correspondingly reduced. Evaluating the detailed expression for $\Delta\beta_0$' given in reference [16] for a 1-mm-wide beam, $w_0=0.6$ mm and $n_{2c}\approx10^{-9}m^2/W^{11}$, the power required for the onset of nonlinear coupling in the waveguide (of which ≈3.4% is carried in the nonlinear medium) is a few milliwatts, in qualitative agreement with experiment. This value is more than one order of magnitude less than that required for the nonlinear guided waves discussed above, and hence that phenomenon is not a complicating factor here.

The optimum coupling efficiency at low powers does not correspond to the optimum coupling efficiency at high powers. In a separate experiment, the coupling was optimized first at low powers and then the power was increased to its final value. The coupling efficiency was then re-optimized by adjusting both the incidence angle and the position of the beam along the prism base, resulting in an increase in efficiency of 45%. Note that this new optimum is always less than if the material response was completely linear, since the cumulative nonlinear phase-shift can never be completely eliminated over the full coupling region. Furthermore, when the power is now decreased, the low power slope is less than for the initial case. These results are all in qualitative agreement with calculations.

The operation of a nonlinear out-coupler is different from the input coupler. In the out-coupling process, the direction of the radiation field is determined by the local guided wave wavevector, and hence it is automatically always synchronized to the guided wave. Reversing the input and output, we found the transmitted intensity remained linear with input power, as expected.

The decrease in coupling efficiency can be reduced by decreasing the beams spot size in conjunction with decreasing the prism-waveguide gap, or by offsetting Δβ to an appropriate value. One feature always obtained is distortion in the field profile; that is, for a Gaussian input beam, the guided wave field is no longer Gaussian along its wavefront. A grating coupler is expected to operate in exactly the same way.

5. Nonlinear Distributed Feedback and Applications

Recently, there has been great progress in making bistable devices [13] and optical logic gates [31] based on intensity-dependent media placed inside Fabry-Perot cavities. Focused plane waves have been used to minimize the power requirements. This is an ideal area for guided wave devices in which high power densities can be maintained over long distances.

An important element required to produce optical bistability is that the nonlinear medium should be placed inside a resonator. In terms of guided waves, this can easily be accomplished with distributed feedback gratings. These can be situated at both ends of the nonlinear medium to act solely as mirrors, or the grating can be fabricated in or on the nonlinear region to perform distributed feedback inside the nonlinear medium.

The linear operational characteristics of gratings are well-known [32]. Because of a grating's reflection properties, two waves are always present inside a grating that we label $a_+(x)$ and $a_-(x)$ for propagation along the +x and −x axes respectively. For a sinusoidal surface grating centered on the plane z=0, which produces a surface corrugation given by $u = u_0\sin(\kappa x)$, and including both the forward and backward traveling waves, the appropriate coupled mode equations including an intensity-dependent wavevector are [33]

$$i\frac{d}{dx}a_+(x) = \Gamma e^{-i\Delta\beta x}a_-(x) + \Delta\beta_0'[a_+(x)^2 + 2a_-(x)^2]a_+(x) \tag{20}$$

$$-i\frac{d}{dx}a_-(x) = \Gamma e^{i\Delta\beta x}a_+(x) + \Delta\beta_0'[2a_+(x)^2 + a_-(x)^2]a_-(x). \tag{21}$$

Here

$$\Gamma = \frac{\omega\epsilon_0}{8}u_0[n_c^2 - n_f^2]\ [E_{+c}(z+0_+).E_{-f}(z+0_-)^*] \tag{22}$$

for a planar waveguide, and

$$\Gamma = \frac{\omega\epsilon_0}{8}u_0\int_{-\infty}^{\infty}[n^2(y,z+0_+) - n^2(y,z+0_-)]\ [E_+(z+0_+).E_-(z+0_-)^*]\ dy \tag{23}$$

for a channel guide with

$$\Delta\beta = 2\beta_0 - \kappa/k_0. \tag{24}$$

These equations have been solved analytically [33,34] in terms of an incident switching power $|a_{sw}(0)|^2$ given by

$$|a_{sw}(0)|^2 = \frac{2\beta}{3\Delta\beta_0'k_0L}. \tag{25}$$

This is typically the power required to obtain switching. Writing the incident and transmitted power as $I = |a_+(0)|^2/|a_{sw}(0)|^2$ and $J = |a_+(L)|^2/|a_{sw}(0)|^2$, then for $\Delta\beta=0$ the grating transmission is given by

$$T = J/I = 2(1 + nd\{2\sqrt{[(\Gamma L)^2 + J^2]};[1 + (J/(\Gamma L))^2]^{-1}\})^{-1} \tag{26}$$

Here nd(u;m) is one of the tabulated Jacobian elliptic functions. As shown in Fig. 5, switching occurs for $|a_+(0)|^2 \approx |a_{sw}(0)|^2$, provided that the feedback parameter is of the order of unity or larger.

Guided wave nonlinear distributed feedback gratings hold considerable promise for low power all-optical devices. Sample calculations for InSb ridge and buried channel waveguides, in which the surface corrugation was

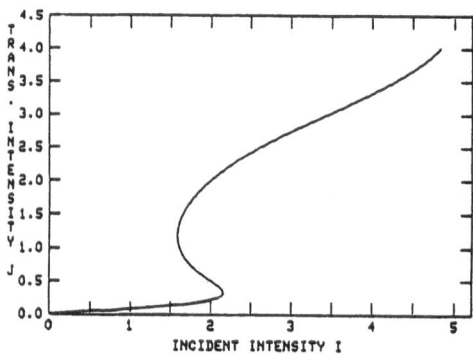

Fig. 5. The calculated guided wave power transmitted by a nonlinear distributed feedback grating versus incident guided wave power normalized to the switching power for $\Gamma L=2$

186

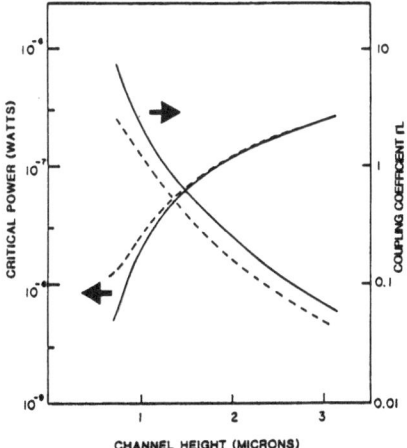

CHANNEL HEIGHT (MICRONS)

Fig. 6. The switching power and the feedback parameter ΓL versus the channel width for a height to width ratio of 0.4 for an InSb channel with air above and Al_2O_3 as the substrate (the solid line is for a ridge waveguide and the dashed line for a buried waveguide)

assumed to be at the channel-air boundary are shown in Fig. 6 [34]. Clearly,bistability should be possible with power levels of a few tens of nanowatts. Polydiacetylene (PTS) ridge waveguides should be capable of switching with power levels of 100's of milliwatts. A host of other devices based on nonlinear gratings are possible [35]. They include optical switching, logic, amplifiers, scanners etc.

No experiments have been reported to date. We are currently fabricating InSb thin film waveguides for this purpose.

6. Degenerate Four Wave Mixing

The key feature that leads to an intensity-dependent refractive index is a nonlinear polarization source field at the same frequency as the incident field. It is also possible to produce such a polarization field by the mixing of three input waves. If two of the input fields propagate in opposite directions, this process is called degenerate four-wave mixing [12]. For any angle of incidence of the third beam in this geometry, the process is always phase-matched,and the fourth wave is generated backwards along the incidence direction of this third beam.

In addition to the usual phase conjugation application of this phenomenon, a number of signal processing operations can potentially be performed in real time in a waveguide geometry [36]. The case shown in Fig. 7a corresponds to convolution. The two input beams being convolved are 1(=a) and 2(=b) and beam 3(=c) is a control beam,whose presence is required to make the process possible via degenerate four-wave mixing. For this case $\theta \simeq 90°$ is preferable. If waves 1 and 2 have pulse envelopes (in time) of the form $U_1(t)$ and $U_2(t')$, then the total radiated signal is of the form

$$U_4(t) = \int U_1(2t-\tau)U_2(\tau) \, d\tau. \tag{27}$$

This corresponds to the convolution of the two input waveforms with a time compression of a factor of two.

Another example is time inversion. This case is illustrated in Fig. 7b. The waveform to be inverted is coupled into a slow mode (TE_m') of the

187

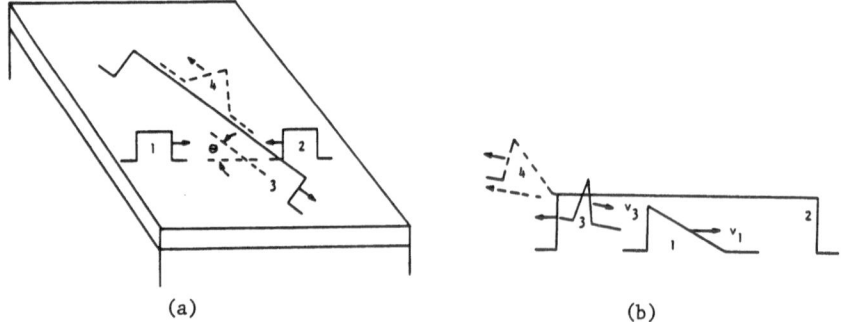

(a)   (b)

Fig. 7. The application of degenerate four wave mixing to (a), the convolution of pulses 1(=a) and 2(=b), and (b), to the time inversion beam 1(=a)

waveguide and a very short pulse (δ-function in time) in a fast mode (TE_m', m<m') overtakes it essentially at a small angle from behind. Hence, the trailing end of the input signal pulse is re-radiated back along the path of the δ-function pulse, and as the short pulse passes through waveform 1, a time-inverted version of the input signal is produced. In this case the control beam is 2 and the radiated signal is beam 4.

The starting point of the analysis is again the appropriate form of the nonlinear polarization field [37]. For three separate input beams labelled a, b and c, the i'th component is

$$P_{\gamma i}(z) = 2c\epsilon_0{}^2 n_\gamma{}^2 n_{2\gamma} \left[\frac{2}{3} E^c{}_{\gamma i}(z) E^a{}_{\gamma j}{}^*(z) E^b{}_{\gamma j}(z) + \frac{1}{3} E^c{}_{\gamma j}(z) E^b{}_{\gamma j}(z) E^a{}_{\gamma i}{}^*(z) \right] (28)$$

in each medium where $n_{2\gamma} \neq 0$. Assuming that beams 1 and 2 propagate along x' (which is oriented at an angle θ to the x-axis), that all four modes have the same mode number, that loss can be included via a complex effective index $\beta - i\beta_I$, and that there is no pump beam depletion,

$$a_a(x') = a_a(0) \, e^{-\beta_I k_0 x'} \qquad \text{and} \qquad (29a)$$

$$a_b(x') = a_b(L') \, e^{-\beta_I k_0 (L'-x')} \qquad (29b)$$

where the beams enter the interaction region at x'=0 and x'=L' respectively. Furthermore, we assume that beam c has the form

$$a_c(x) = a_c(L) \, e^{-\beta_I k_0 (L-x)} \qquad (29c)$$

where beam c is injected at the point x=L. From coupled mode theory, the fourth wave amplitude is given by

$$\frac{d}{dx} a_d(x) + \beta_I k_0 a_d(x) = \frac{k_0 c}{4i} \int P_i(z) E^d{}_i{}^*(z) \, dz \, a_a(0) a_b(L') a_c(L) e^{-\beta_I k_0 L' - \beta_I k_0 (L-x)}. \qquad (30)$$

We now group the details of the integration over the depth dependence of the fields into a single term A,

$$A = \frac{2k_0 c^2 \epsilon_0{}^2}{3i} (2+\cos^2\theta) \int_{-\infty}^{\infty} n_\gamma(z)^2 n_{2\gamma}(z) \, E^a{}_{\gamma y}(z) E^b{}_{\gamma y}(z)^* E^c{}_{\gamma y}(z) E^d{}_{\gamma y}(z)^* dz \qquad (31)$$

188

for all TE waves. The solution to (30) is simply

$$a_d(x) = \frac{Ae^{-(\beta_I k_0 L' + \beta_I k_0 L)}}{2\beta_I k_0} [e^{\beta_I k_0 x} - e^{-\beta_I k_0 x}] \, a_a(0)a_b(L')a_c(L) \tag{32}$$

In this calculation we are primarily interested in the fourth wave leaving the interaction region at x=L and hence

$$a_d(L) = \frac{A[e^{-\beta_I k_0 L'} - e^{-(\beta_I k_0 L' + 2\beta_I k_0 L)}]}{2\beta_I k_0} \, a_a(0)a_b(L')a_c(L). \tag{33}$$

Typically, the beam d signal power is of experimental interest. Writing $|a_a(0)|^2 = P_a/H'$, $|a_b(L')|^2 = P_b/H'$, $|a_c(L)|^2 = P_c/H$ and $|a_d(L)|^2 = P_d/H$ where beams a and b are H' wide and beams c and d are H wide,

$$P_d = |A|^2 (\frac{L}{H'})^2 \, G^2 \, P_a P_b P_c , \tag{34}$$

where the effect of attenuation is included in

$$G = \frac{e^{-\beta_I k_0 L'} - e^{-(\beta_I k_0 L' + 2\beta_I k_0 L)}}{2\beta_I k_0 L} \tag{35}$$

and $D^{NL} = |A|^2$. For $2\beta_I k_0 L \ll 1$, i.e., negligible attenuation over the interaction region, $G \to 1$. For $L = L'$, $m = m'$ and all waves either TE or TM polarized which is the usual case, the optimum interaction distance is $L_{opt} = 1.099/2\beta_I k_0$. Under these conditions, the optimum signal power is

$$P_d = D^{NL} \frac{1}{27H^2 k_0^2 \beta_I^2} \, P_a P_b P_c . \tag{36}$$

One can estimate the input powers required for strong conversion of the input to output signals. From (34), this power is given approximately by $P_a \mathrm{\iota} [D^{NL}]^{-1/2}$. In this limit, however, it is necessary to take incident beam depletion into account, which leads to a series of coupled mode equations between the amplitudes of the various beams. This case has not been considered yet for guided waves.

Experiments were carried out using liquid CS_2 as a nonlinear cladding material [38]. Figure 8 shows the sample and beam interaction geometry used. The waveguide consists of a ≈ 1-μm-thick film of Corning 7059 glass deposited on a BK-7 substrate. Three high index coupling prisms are arranged on the free film surface so that the three input guided wave pump and probe beams would intersect. A small quantity of CS_2, which has a large (relative to glass) third order nonlinearity $\chi^{(3)} \approx 2 \times 10^{-20}$ m^2/V^2, was used as a nonlinear cover medium over the part of the waveguide where the beams intersect. The liquid was held in place by a hollowed out glass cell whose bottom surface was well-polished and optically contacted to the thin film surface. After careful alignment of the three coupled input beams, the glass cell is inserted, gently pressed onto the waveguide via a set screw, and filled from the top with CS_2. Thus, the evanescent tails of the guided beams in the CS_2 generate the nonlinear source polarization. The efficient coupling of three independent beams with prisms was found to be very difficult to achieve, and resulted in reduced coupling efficiencies, of the order of 1%.

A passively Q-switched, frequency doubled ($\lambda = 0.53$ μm) Nd$^+$:YAG laser was used in the experiments. Due to poor coupling efficiencies (under 1%),

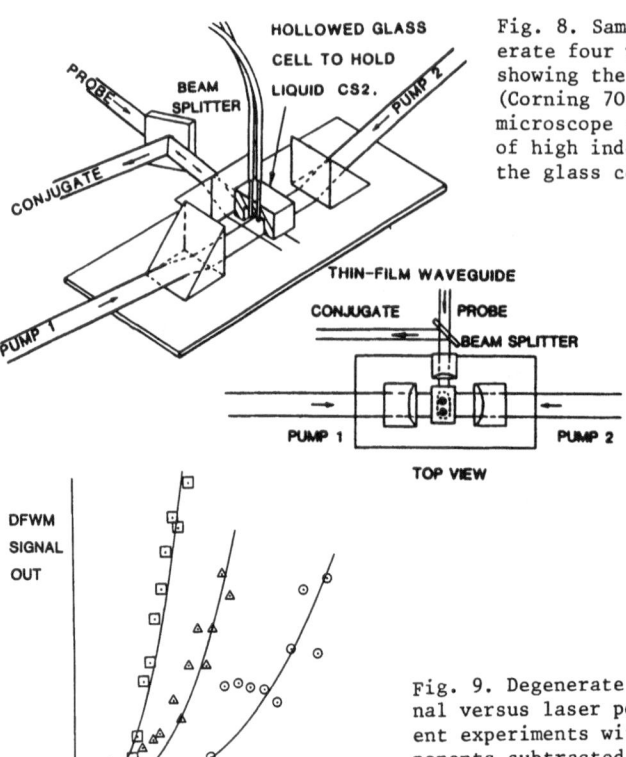

Fig. 8. Sample and coupling degenerate four wave mixing geometry showing the thin film waveguide (Corning 7059 sputtered onto a glass microscope slide), the arrangement of high index coupling prisms, and the glass cell for holding CS_2

Fig. 9. Degenerate four wave mixing signal versus laser power for three different experiments with the background components subtracted out (the curves drawn represent the best cubic fit)

peak power densities achieved within the waveguide were only on the order of 100 MW/cm² in each beam, and of the order of 6 MW/cm² in the CS_2 which carried only 6% of the guided wave power. Despite giving a smaller interaction region than a more colinear geometry would provide, a 90° probe angle (relative to the pump beams) was selected to minimize background contamination of the signal due to stray scattered light from the pump beams.

Results are shown in Fig. 9 and the cubic dependence of the DFWM signal on incident power is clear. Each curve represents a different trial taken under different experimental conditions; that is, different coupling efficiencies and detector gain. The background signal due to stray light was first subtracted off.

The magnitude of the conjugate signal was calculated as outlined above. Estimating P_d for our experimental conditions (detector sensitivity and coupling efficiency), we obtained a value of 0.01 pJ per pulse for a guided wave input in the nonlinear medium of 3 μJ per input beam. This is in reasonable agreement with the measured value of 0.005 pJ based on the dominant reorientational contribution to the nonlinear susceptibility of CS_2. The thermal contribution to the nonlinearity, estimated from the CS_2 absorption coefficient, its thermal capacity, and temperature-dependence of refractive index, was found to be negligible.

No measureable signal was obtained in the absence of the CS_2. This confirmed that the nonlinear mixing occurred in the nonlinear cladding medium, in agreement with calculations that included contributions from the glass film as well as the CS_2.

The major difficulties that we encountered were related to the relatively low DFWM cross-section obtained with CS_2 used in a cladding geometry, of the order of 10^{-9}. For the available laser powers in the waveguide, this resulted in stray light signal levels comparable to the DFWM signal, especially emanating from the probe-signal prism coupling region. We expect that the use of more highly nonlinear materials, such as PTS for the film itself, will lead to useful DFWM signals for all-optical signal processing.

7. Surface Coherent Raman Spectroscopy

CARS, or Coherent Anti-Stokes Raman Spectroscopy is a nonlinear spectroscopic tool for probing the vibrational spectra of a material. It involves the nonlinear mixing of two input laser beams of frequency ω_a and ω_b, one of which is tuneable. The pertinent guided wave geometry is shown in Fig. 10. When the two guided wave beams are incident onto a molecule such that the frequency difference $\omega_a-\omega_b$ corresponds to one of the vibrational frequencies of the molecule, i.e., $\omega_a-\omega_b \approx \omega_p$, the molecular vibration is resonantly excited. If $\omega_a-\omega_b \neq \omega_p$, the molecular vibration is not excited. For incident fields of the form given by (2), phased arrays of molecules are coherently excited with a spatial and temporal dependence given by $\exp[i(\omega_a-\omega_b)t - i(\mathbf{k}_a-\mathbf{k}_b)\cdot\mathbf{r}]$. These coherently excited molecular vibrations now scatter light from one of the beams in the usual Raman scattering process , producing nonlinear polarization fields at the frequencies $2\omega_a-\omega_b$ (Coherent Anti-Stokes Raman Scattering, CARS) and $2\omega_b-\omega_a$ (Coherent Stokes Raman Scattering, CSRS), and the wavevectors $2\mathbf{k}_a-\mathbf{k}_b$ and $2\mathbf{k}_b-\mathbf{k}_a$ respectively. Under the right conditions, one of these polarization fields can radiate in a phase-matched way into a waveguide mode. For example, for CARS in a waveguide geometry, phase matching occurs if the incident field wavevectors are arranged so that $\mathbf{k}_d=2\mathbf{k}_a-\mathbf{k}_b$ where $\mathbf{k}_d$ is the wavevector of a guided wave. This phase-matching condition is crucial for obtaining strong signals, since the scattered field amplitude grows linearly throughout the interaction volume only if phase-matching is enforced.

The pertinent nonlinear polarization is given in reference [39]. The angle between the beams a and b required for phase-matching is small,

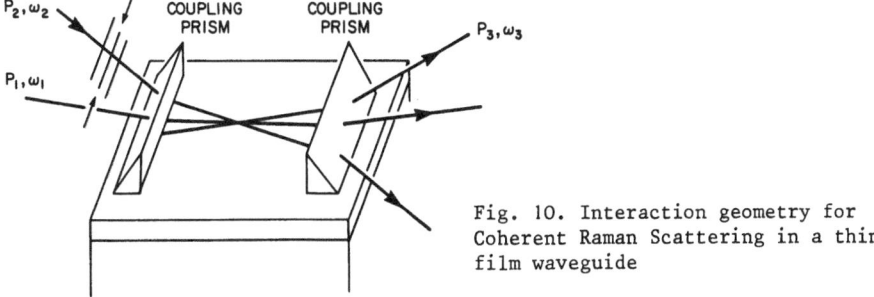

Fig. 10. Interaction geometry for Coherent Raman Scattering in a thin film waveguide

191

usually at most a few degrees. For the CARS case using TE waves in optically isotropic media

$$P_y(z) = \varepsilon_0 \chi^{(3)}_{zzzz}(-\omega_d, -\omega_b, \omega_a, \omega_a) E^a_y(z)^2 E^b_y(z)^*$$ (37)

Using the definition for the fields in equation 16, and applying coupled mode theory

$$\frac{d}{dx}a_d(x) + \beta_{dI}k_0 a_d(x) = \frac{k_0 c \varepsilon_0}{4i} F \; C_a^2 C_b \; a_a(0)^2 a_b(0) \; e^{-(2\beta_{aI} + \beta_{bI})k_0 x}$$ (38)

where

$$F = \int_\infty^\infty \chi^{(3)}_{zzzz}(z) \; f_a^2(z) f_b(z) f_c(z) \; dz.$$ (31)

The parameter F defines the appropriate overlap integral whose magnitude will vary with the combination of guided wave modes used. As usual, the signal power is important, and this is usually written in the form

$$P_d = D^{NL}\left[\frac{L}{H}\right]^2 P_a^2 P_b \; ,$$ (32)

where all the details of the guided wave aspects of the interaction are summarized in D^{NL}.

The key to the above results is that P_d is proportional to $|\chi^{((3))}|^2$. The spectroscopy aspect is clearly contained in the frequency response of the susceptibility term. In general,

$$\chi^{(3)} = \chi_b^{(3)} + \sum_p \chi_{rp}^{(3)} \; \frac{\Gamma_p}{\omega_a - \omega_b - \omega_p - i\Gamma_p} \; ,$$ (33)

where the summation over p is the summation over the molecular vibrational modes of the sample characterized by frequency ω_p, linewidth Γ_p and resonant susceptibility $\chi_{rp}^{(3)}$. Here $\chi_b^{(3)}$ is the background susceptibility due to electronic degrees of freedom of the molecules. Clearly, as the frequency difference is tuned through the characteristic vibrational frequencies, resonant enhancement is obtained in $\chi^{(3)}$ and the CARS signal exhibits a maximum.

Numerical calculations for thin films based on realistic materials indicate that the signal power on a strong resonance can be very large. For example, we consider a 0.35-μm-thick polystyrene film deposited on a glass substrate, and the 1000-cm⁻¹ ring vibration. For laser pulse energies of less than 0.1 mJ in pulses of ≈15-ns duration, efficiencies of ≈0.5% are predicted for the conversion of the incident beams into the CARS signal. Such laser energies are easily available making this approach very promising for nonlinear spectroscopy.

The initial experiments were carried out on 2-μm-thick polystyrene waveguides using two tuneable dye lasers with 100-ps-long pulses [40]. The power densities in the two beams were 30 and 60 MW/cm², which corresponded to pulse energies of hundreds of nanojoules. The spectrum obtained is shown in Fig. 11, and at the peak, the conversion efficiency was 0.2%, in excellent agreement with the theoretical calculations.

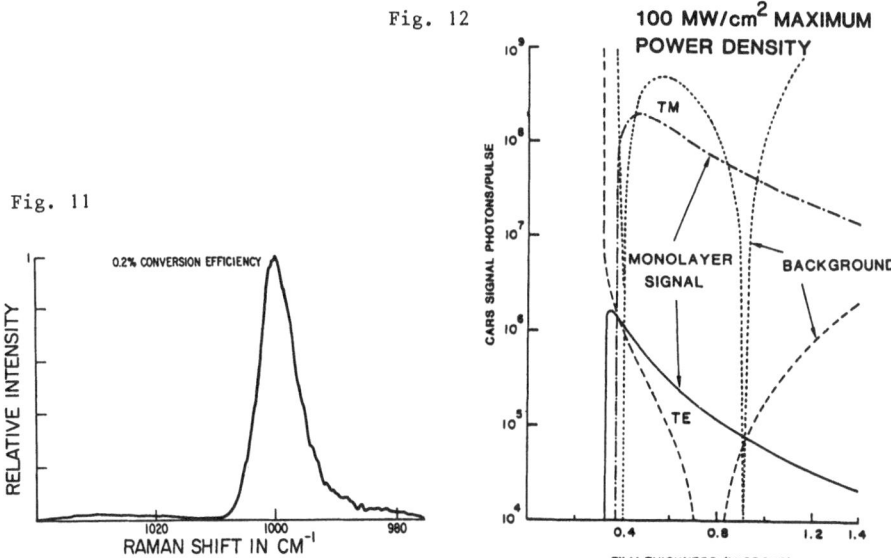

Fig. 11. CSRS spectrum for the ring vibration of a polystyrene thin film waveguide

Fig. 12. Resonant and background CARS signals in photons per pulse calculated for a benzene monolayer on a Nb_2O_5-SiO_2 waveguide versus Nb_2O_5 film thickness. For TE waves: (solid line) CARS signal, (long-dashed line) background. For TM waves: (dotted-dashed line) CARS signal, (short-dashed line) background. The modes used were $TE_2(\omega_a)$, $TE_1(\omega_b)$, and $TE_3(\omega_a)$; or $TM_2(\omega_a)$, $TM_1(\omega_b)$, and $TM_2(\omega_d)$

The remarkable efficiency of this process suggests applications to the investigation of monolayers deposited on film surfaces [39]. The problem is that the background term for the film can lead to a large signal that can mask the desired monolayer signal, that is, under normal conditions the guiding film can be thousands of angstroms thick where—as a monolayer has a thickness of, say, 10 Å. The solution to this background problem lies in the coherent nature of this process, and the advantages of using guided waves. If at least one of the modes has m>0, interference effects occur in F, and hence the film contribution can be minimized, leaving the monolayer term as the dominant contribution. Such a calculation is shown in Fig. 12, which shows that the film contribution can be negligible over a range of thickness, providing an appropriate set of modes is used. Experiments are currently in progress for detecting monolayers on the surface of a waveguide, and the preliminary results indicate that the background term can in fact be reduced substantially.

8. Summary

Integrated optics offers an ideal configuration for nonlinear optics depending on the third order susceptibility. We have described a number of new experiments which utilize this phenomenon. In particular, waveguide versions of plane wave nonlinear optical phenomena, such as four wave mixing and coherent Raman spectroscopy, have been demonstrated. The four wave mixing experiment is just a first step towards all-optical, real time signal processing operations such as convolution and time inversion. The CARS technique holds considerable promise for studying the nature and orientation of monolayers on a film surface. Not achieved yet experimentally, but imminent, are optical bistability and switching based on distributed feedback in nonlinear waveguides.

We have also demonstrated two phenomena based on intensity- dependent refractive indices which are unique to waveguides and have no analog in plane wave nonlinear optics, both of which exhibit optical limiting action. The first is the decrease in coupling efficiency, with increasing power obtained with distributed couplers such as prisms or gratings. We have also demonstrated the existence of nonlinear guided wave modes, whose field distributions and effective indices exhibit strong power dependences.

This research was supported by the National Science Foundation (DMR-8300599) and (ECS-8304749,-8312845,-8117483).

References

1. Y.R. Shen: The Principles of Nonlinear Optics (J. Wiley, New York, 1984)
2. D,B. Anderson and J. T. Boyd: Appl. Phys. Lett. 19, 266 (1971)
3. Y. Suematsu, Y. Sasaki and K. Shibata: Appl. Phys. Lett. 23, 137 (1973)
4. J.P. van der Ziel, R.C. Miller, R.A. Logan, W.A. Nordland Jr. and R.M. Mikulyak: Appl. Phys. Lett. 25, 238 (1974)
5. A.T. Reutov and P.P. Tarashchenko: Opt. Spectrosc. 37, 447 (1974)
6. W. Sohler and H. Suche: Appl. Phys. Lett. 33, 518 (1978)
7. M. De Micheli, J. Botineau, S. Neveu, P. Sibillot and D. B. Ostrowsky: Opt. Lett. 8, 116 (1983)
8. W. Sohler and H. Suche: 1980, Appl. Phys. Lett. 37, 255 (1980)
9. W. Sohler and H. Suche: Proc. SPIE Vol 408, in press (1983)
10. R. Normandin and G.I. Stegeman: Appl. Phys. Lett. 36, 253 (1980)
11. R. Normandin and G.I. Stegeman: Appl. Phys. Lett. 40, 759 (1982)
12. D.M. Pepper: Opt. Eng. 21, 156 (1982)
13. D.A.B. Miller, S.D. Smith and C.T. Seaton: IEEE J. Quant. Electron. QE-17, 312 (1981
14. M.D. Levenson: Introduction To Nonlinear Laser Spectroscopy, (Academic Press, New York 1982
15. P.D. Maker and R. Terhune: Phys. Rev. 137, A801 (1964)
16. G.I. Stegeman: IEEE J. Quant. Electron. QE-18, 1610 (1982)
17. N.N. Akhmediev: Sov. Phys. JETP 56, 299 (1982)
18. A.A. Maradudin: "Nonlinear Surface Electromagnetic Waves", in Proc. Second Int. School on Condensed Matter Physics, Varna, Bulgaria (World Scientific Publ., Singapore 1983)
19. D.J. Robbins: Optics Comm. 47, 309 (1983)
20. F. Lederer, U. Langbein and H.-E. Ponath: Appl. Phys. B 31, 69 (1983)
21. A.D. Boardman and P. Egan: J. de Phys. Colloque, C5,291 (1984)
22. G.I. Stegeman, C.T. Seaton, J. Chilwell and S.D. Smith: Appl. Phys. Lett. 44, 830 (1984)
23. C.T. Seaton, J.D. Valera, R.L. Shoemaker, G.I. Stegeman, J. Chilwell and S.D. Smith: Appl. Phys. Lett. 45, 1162 (1984)
24. A.E. Kaplan: Sov. Phys. JETP 45, 896 (1977)
25. C.T. Seaton, J.D. Valera, B. Svenson and G.I. Stegeman: Opt. Lett. 10, in press March (1985)
26. H. Vach, C.T. Seaton, G.I. Stegeman and I.C. Khoo: Opt. Lett. 9, 238 (1984)
27. C. Liao and G.I. Stegeman: Appl. Phys. Lett. 44, 164 (1984)

28. C. Liao, G.I. Stegeman, C.T. Seaton, R.L. Shoemaker, J.D. Valera and H.G. Winful: J. Opt. Soc. Am. B, 2, in press (1985)

29. R. Ulrich: J. Opt. Soc. Am. 60, 1337 (1970)

30. J.D. Valera, C.T. Seaton, G.I. Stegeman, R.L. Shoemaker, Xu Mai and C. Liao: Appl. Phys. Lett. 45, 1013 (1984)

31. C.T. Seaton, S.D. Smith, F.A.P. Tooley, M.E. Prise and M.R. Taghizadeh: Appl. Phys. Lett. 42, 131 (1983)

32. H. Kogelnik: 1975, in Integrated Optics, Vol 7 of Topics in Applied Physics, T. Tamir, ed., (Springer Verlag, Berlin 1975)

33. H.G. Winful, J.H. Marburger and E. Garmire: Appl. Phys. Lett. 35, 379 (1979)

34. G.I. Stegeman, C. Liao, and H.G. Winful: "Distributed feedback bistability in channel waveguides," in Optical Bistability II, edited by C.M. Bowden, H.M. Gibbs and S.L.McCall (Plenum Press, New York 1984) p389

35. H.G. Winful and G.I. Stegeman: "Applications of nonlinear periodic structures in guided wave optics", SPIE Proc. of First Int. Conf. on Integrated Optical Eng., in press (1985)

36. G.I. Stegeman: J. Opt. Comm. 4, 20 (1983)

37. C. Karaguleff and G.I. Stegeman: IEEE J. Quant. Electron. QE-20, 716 (1984)

38. C. Karaguleff, G.I. Stegeman, R. Zanoni and C.T. Seaton: Appl. Phys. Lett. 46, in press (1985)

39. G.I. Stegeman, R. Fortenberry, C. Karaguleff, R. Moshrefzadeh, W.M. Hetherington III, and N.E. Van Wychk: Opt. Lett. 8, 295 (1983)

40. W.M. Hetherington III, N.E. Van Wyck, E.W. Koening, G.I. Stegeman and R.M. Fortenberry: Opt. Lett. 9, 88 (1984)

Optimized Structure of Ti:LiNbO$_3$ Channel Waveguides for Optical Parametric Oscillators

G.P. Bava and I. Montrosset

Dipartimento di Elettronica, Politecnico di Torino, C.so Duca degli Abruzzi 24
I-10129 Torino, Italy

W. Sohler and H. Suche

Universität-Gesamthochschule Paderborn, Angewandte Physik, Postfach 1621
D-4790 Paderborn, Fed. Rep. of Germany

The threshold power of a Ti:LiNbO$_3$ integrated optical parametric oscillator was evaluated as function of the various waveguide parameters to find a structure optimized for strongest nonlinear interaction.

1. Introduction

Recently, the first integrated optical (Ti:LiNbO$_3$) parametric oscillators (IOPO) were reported [1]. Tunable frequency conversion from the visible ($\lambda_p \cong 0.6$ µm) to the near infrared (0.9 µm $< \lambda_s$, $\lambda_i <$ 1.7 µm) was demonstrated. The minimum threshold pump power was 2.8 W, the maximum power conversion efficiency 14%. Despite these attractive features, the IOPO has the potential to be improved further. In particular, the threshold power should be lower for a practical device. To achieve this goal, research and development have to follow mainly two lines. First, it is necessary to reduce the waveguide losses further to improve the resonator finesse. This is essentially a technological problem. Second, the waveguide profile has to be optimized to get maximum nonlinear interaction of pump-, signal- and idler- modes. This requires an extensive analysis of the threshold pump power with respect to the various parameters characterizing the waveguide. As diffusion coefficients and relation between increase of index of refraction and Ti - concentration are known from experiments [2], this analysis can be done numerically. It is the aim of this paper to develop an efficient computer program for a fast evaluation of the mode field distributions (Sec.2), of the threshold power (Sec.3) and of the optimized waveguide structure (Sec.4).

2. Waveguide model and field evaluation.

The Ti^{4+} -ion concentration c of a $Ti:LiNbO_3$ optical waveguide (Fig.1) can be given by the following expression [3]:

$$c(z,y)= c_0 f(y) g(z) \qquad (1)$$

where

$$f(y) = \exp(-y^2 /D^2) \quad ,$$

$$g(z) = 0.5 \{erf[A(1-B)] + erf[A(1+B)]\} \quad ,$$

and $A = W/2D$, $B = 2 z/W$, W: width of the Ti-stripe before diffusion, D: diffusion depth, $c_0 g(0)$: maximum Ti concentration. The indiffused ions produce an increase of the ordinary (Δn_o) and extraordinary (Δn_e) indices of refraction [2], which can be represented as follows:

$$\Delta n_e (y,z) = E\ c(y,z) \qquad (2a)$$

$$\Delta n_o (y,z) = [\ F\ c(y,z)\ ]^{\gamma} \qquad (2b)$$

with: $E= 1.14\ 10^{-23}$ cm^3, $F = 0.13\ 10^{-24}$ cm^3, $\gamma = 0.55$. Equations (2a-b) characterize the optical waveguide. To evaluate its optical modes, Δn_e and Δn_o depth dependences were approximated by means of a function of the hyperbolic tangent [4]. These expressions were inserted in the wave equation and dispersion was taken into account. Assuming

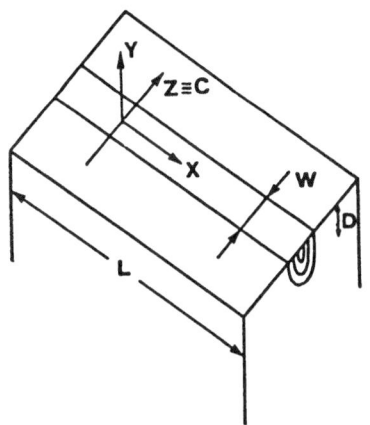

Fig.1 Geometrical waveguide parameters of the parametric oscillator.

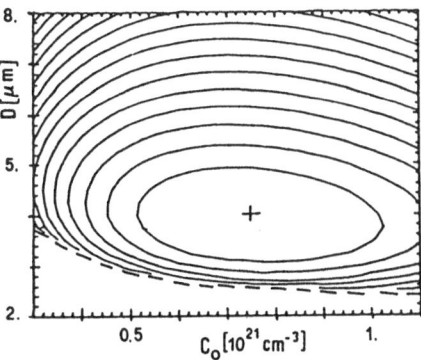

Fig.2 Constant P_t curves, equidistant 0.5 mW, in D-c_0 plane with W=7. µm, * P_t min., lower P_t curve 15. mW, - - - idler cutoff.

the quasi-TE or -TM mode approximation to hold, the wave equation can be solved, using the "effective index method" [5], in two steps. First, the wave equation is integrated along the y-axis with fixed z as parameter. With our approximations, an analytic espression for the effective index profile along the z-axis is obtained which is used in the second step to integrate the remaining one-dimensional wave equation. This profile is approximated further by means of a function of either the hyperbolic or the trigonometric tangent [4], which results in a very fast computation of the mode effective index and of the field distributions in terms of Gegenbauer polynomials [4,6]. These results were carefully checked and compared for some sets of parameters with the results obtained by numerical techniques; the coincidence is satisfactory [4].

3. Calculation of the Threshold Power.
In the following analysis of parametric oscillator operation, the relation $\omega_p = \omega_s + \omega_i$ between pump, signal and idler angular frequencies must be satisfied, together with the phase matching condition $\beta_p = \beta_s + \beta_i$. In order to easily achieve the phase matching, a quasi-TM_{oo} mode has been selected for signal and idler, and a quasi-TE_{oo} mode for the pump. To calculate the threshold power of the IOPO only doubly resonant structures (i.e. resonant for both signal and idler waves) are considered. The mirrors forming the oscillator cavity are supposed to be transparent at the pump wavelength . If , moreover , the signal and idler mode attenuation constants are assumed to be identical ($\alpha_s = \alpha_i = \alpha$), the following explicit expression can be found for the threshold power P_t :

$$P_t = (\omega_s\ \omega_i\ k^2\ )^{-1}\{ \alpha\ \ln(M+\sqrt{M^2-1}\ )/[1-\exp(-\alpha_p L)]\}^2$$
where:
$$M = \exp(2\alpha L)\ [R_s\ R_i\ \exp(-4\alpha L)+1]/(R_s + R_i\ )$$

$$R_s = |\ r_{s1}\ r_{s2}\ |\ ,\ \ R_i = |\ r_{i1}\ r_{i2}\ |$$
$$k \simeq (d_{32}\ /2)\ \iint\limits_y E_y^{\omega_s}\ (y,z)\ E_y^{\omega_i}\ (y,z)\ E_z^{\omega_p\ *}(y,z)\ dy\ dz,$$

and L is the resonator length; r are the mirror field reflectivities for signal and idler at the two end faces (1 and 2); d_{32} is the nonlinear coefficient of LiNbO$_3$ (=5.35 10**-23 As/V^2); α_p is the attenuation constant of

the pump mode. The above expression of P_t is factorized in
two independent terms: the first depends on the overlap
integral of the modal field distributions, on the nonlinear
coefficient and on the operation frequencies. This term,
in particular the coefficient k , can be maximized by an
appropriate waveguide structure. The second term expresses
the dependence of the threshold power on α, α_p, R_s, R_i and
L ; it is independent of the waveguide structure and can
be easily evaluated to study the influence of these
parameters. It is found, for instance, that an optimum L
exists for assigned values of the other parameters.

4. Optimization of the Waveguide Structure.
As pointed out above, the waveguide structure essentially
determines the oscillator threshold power by the overlap
integral of the normalized field distributions of the
interacting modes. The operating angular frequencies
determine the crystal temperature T to achieve phase
matching (200 < T < 270 ^{O}C). We assumed at this state of
the calculations operation with a pump wavelength of
λ_p =595 nm and signal and idler wavelengths λ_s =1071.5 nm,
λ_i =1338 nm respectively (not too far from degeneration).
To find the parameters c_O, W and D of the optimized
waveguide structure, we calculated the threshold power for
constant W as function of c_O and D. To do this, typical
resonator data were taken from our experiments :
α =0.3 dB/cm, α_p =0.1 dB/cm, R_s =R_i =0.98, L=48.mm. Maps
of constant P_t were plotted in the D-c_O plane (Fig.2).
Typically, a broad minimum of P_t appears and it becomes
flatter as W increases. The optimized waveguide structure
is found by tracking the minima of P_t for various W. The
result is presented in Fig.3 : the optimized waveguide
has: W=7. μm, D=4. μm and a maximum Ti concentration of
5.8 10^{20} cm^{-3}. In these conditions, the threshold power
turns out to be 14.9 mW. Integration over the
concentration profile yields the number of Ti-ions and
thereby the Ti-layer thickness on the LiNbO$_3$ substrate
before diffusion. Furthermore,the tuning characteristics
of this optimized integrated optical parametric oscillator
were evaluated assuming fixed pump wavelength or fixed
temperature like in Fig.4. The threshold behavior was
calculated as well,and is presented in the same figure.

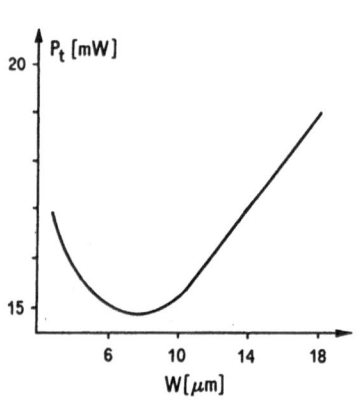

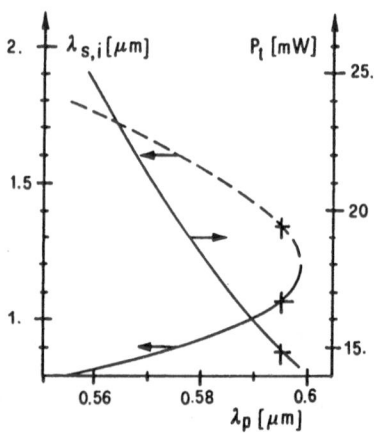

Fig.3 Minimum P_t ,in D-c plane, as a function of W.

Fig.4 Tuning characteristics and P_t for the optimized waveguide as a function of λ_p; + optimum condition.

5. Conclusions.

Using the approximations presented in Sec.2, we have developed an efficient computer program to calculate the threshold power of an integrated optical parametric oscillator. It was possible to perform the extensive computations, needed to evaluate an optimized waveguide structure for nonlinear interactions, in a reasonable (CPU) time. Such a structure is necessary to construct integrated optical parametric oscillators of minimum threshold power. From the calculations (and from our latest experiments) it is evident that parametric oscillators can be pumped by semiconductor laser one day.

Acknowledgements.
Two of the authors (G.P.B., I.M.) gratefully acknowledge Mr. A.Milone,who developed a first version of the computer code. We thank the Italian Consiglio Nazionale delle Ricerche and the Deutsche Forschungsgemeinschaft for financial support of this common project.

References
1 W.Sohler,H.Suche:Third Int. Conf. on Int. Optics and
   Optical Fibres Comm.,1981,San Francisco,paper WB1,p.89.
   W.Sohler, H.Suche: "Frequency Conversion in Ti:LiNbO$_3$
   Optical Waveguides",SPIE Proc.,408,163,(1983).

H.Suche:Theses, Dortmund University, (1981).

2 H.Ludtke, W.Sohler, H.Suche: "Characterization of
Ti:LiNbO$_3$ Waveguides", in Digest of Workshop on
Integrated Optics (R.T.Kersten,R.Ulrich Ed.),1980,p.122.

3 W.K.Burns,P.H.Klein,E.J.West:J.Appl.Phys.,50,6175(1979)

4 G.P.Bava, I.Montrosset, E.Strake: to be published.

5 G.B.Hocker, W.K.Burns: Appl. Optics,16,113,(1977).

6 J.K.Butler,D.Botez:IEEE J. of Q.E.,QE-18, 952, (1981).

Single Mode Channel Waveguide Polarizer on LiNbO$_3$

D. Eberhard and H. Bülow

Universität Dortmund, Lehrstuhl für Hochfrequenztechnik, Postfach 500500
D-4600 Dortmund 50, Fed. Rep. of Germany

Single mode stripe waveguides with a dielectric/metal overlay operate as
absorption polarizers. For Ti:LiNbO$_3$ TM/TE differential losses of 100 dB/cm
are measured.

1. Introduction

Integrated-optic waveguide polarizers are important components in integrated-optic devices, first of all for sensor technology. A variety of solutions for waveguide polarizers has been demonstrated based on the following principles: metal-clad waveguides /1/, proton exchanged waveguides
/2,3/, birefringing superstrates /4/, spatial separation of polarization
states /5/. Here, stripe waveguide polarizers with a dielectric/metal
overlay are reported. These polarizers are absorptive and do not lead to
uncontrolled substrate light. The absorption of TM-polarized waveguide
modes is due to a coupling to surface plasmons in the metal layer. This
coupling is strongly enhanced by suitable dielectric buffer layer.

2. Principle

The TM-TE differential loss of dielectric/metal clad integrated-optic waveguide polarizers depends on the refractive index n_B of the buffer layer,
the buffer layer thickness d_B, the choice of the metal and the waveguide
parameters. To get a general idea of these dependences, the TM and TE losses
were calculated for dielectric/metal clad Ti:LiNbO$_3$ slab waveguides. The
Gaussian index profile with a diffusion depth x_D is approximated by a multilayer staircase function. The calculations are performed by generalizing
the known transverse resonance technique to a multilayer system, a complex impedance is used for the metal layer taking account of its complex
refractive index. Gold, silver, aluminium, platinum are examined as metals,
the optical data are taken from /6/; silicon dioxide (SiO$_2$, n_B = 1.46),and
silicon nitride (Si$_3$N$_4$, n_B = 1.9) are investigated as buffer materials. The
calculations are for 633 nm wavelength, and the Ti:LiNbO$_3$ waveguide parameters are x_D = 1 µm and a maximum index change $\Delta n/n$ = 0.01. Pt-layers

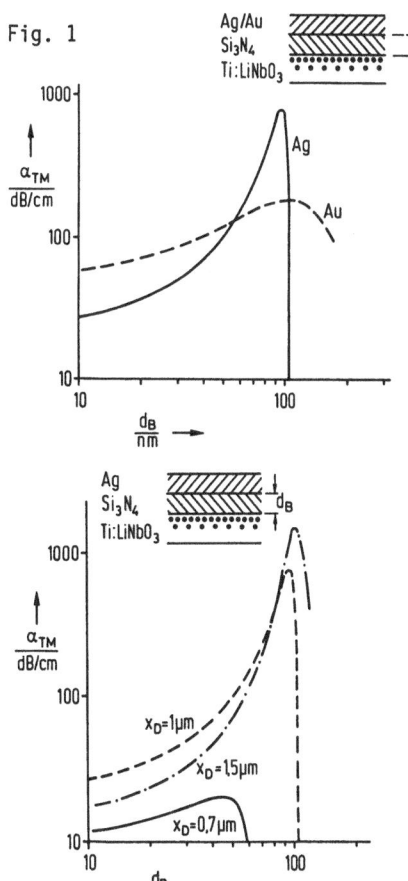

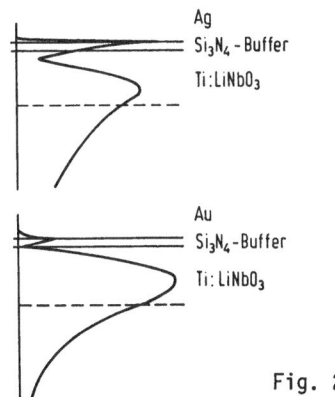

Fig. 1
Calculated TM-losses of a Ti:LiNbO₃ slab waveguide with Ag/Si₃N₄ and Au/Si₃N₄ overlay versus buffer thickness d_B

Fig. 2
Calculated field distribution of a Ti:LiNbO₃ slab waveguide with Ag/Si₃N₄ and Au/Si₃N₄ overlay for an optimum buffer thickness $d_B \simeq 100$ nm

Fig. 3
Calculated TM-losses of Ti:LiNbO₃ slab waveguides with Ag/Si₃N₄ overlay and different diffusion depths x_D versus buffer thickness d_B

yield high TE-losses. Al-layers require extremely thin (∿ 1 nm) buffer layers. Therefore, only Au- and Ag-layers are considered. Some results of the computer analysis are shown in Fig. 1 to 3. In Fig. 1 the TM absorption loss is demonstrated versus the Si_3N_4-buffer thickness d_B for thick silver layers (solid line) and thick fold layers (dashed line). A SiO_2 buffer layer results in a similar diagram, but the optimum buffer thickness is about 5 times smaller than in case of Si_3N_4 leading to difficulties with thickness control. Therefore only Si_3N_4-buffer layers are investigated. The TE-losses are about 1 dB/cm in all cases. Figure 2 shows the calculated field distributions for an optimum Si_3N_4 buffer thickness. It explains the different maximum absorption losses. If silver is used as a metal, the phase of the plasmon wave at the metal surface is matched to the slab waveguide mode leading to a high coupling efficiency and strong TM absorption. No phase matching is possible if gold is utilized. The field distribution within the waveguide is only slightly changed compared to an undisturbed slab waveguide. In Fig. 3 the TM loss is plotted versus Si_3N_4-buffer thickness for a polarizer with Si_3N_4/Ag overlay. The waveguide parameters are $\Delta n = 0.022$ and the diffusion depths of the Gaussian profile are $x_D = 0.7$, 1.0, and 1.5 µm. One recognizes that TM polarized waveguide modes which are weakly guided show lower losses. Furthermore, the optimum buffer thickness decreases for smaller diffusion depths.

3. Experimental

The integrated-optic stripe waveguide polarizers are fabricated by the indiffusion of a set of parallel titanium stripes into X-cut $LiNbO_3$. The titanium stripe width is 2 µm, the thickness 30 nm to 35 nm to achieve monomode waveguides at HeNe-wavelength (633 nm). After indiffusion at 1000°C for 8 h in a water vapor-enriched oxygen atmosphere, the substrates are sputtered with Si_3N_4 in such a way that the buffer thickness increases from one stripe waveguide to the next one. Finally, the device is covered by a 100 nm thick gold or silver layer with 1 mm length perpendicular to the waveguides. Experimental results of the TM losses are demonstrated in Fig. 4.

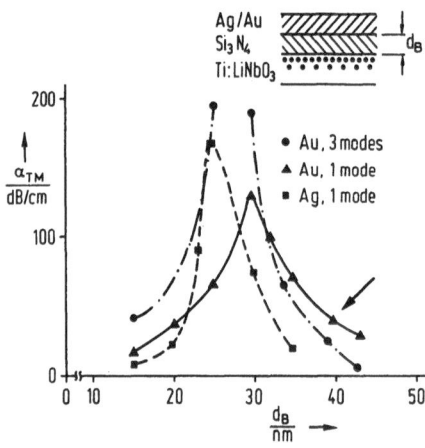

Fig. 4
Measured TM-losses for single-mode stripe waveguides with Ag/Si$_3$N$_4$ and Au/Si$_3$N$_4$ overlays compared to a 3-mode stripe waveguide with Au/Si$_3$N$_4$ overlay

The excess TE-losses due to the metal clad are about 1 - 2 dB/cm. In agreement with the theoretical calculations, silver yields a better TM absorption and a lower optimum buffer thickness than gold. Figure 4 contains in addition results for a stripe waveguide that carries 3 lateral modes (all modes are excited). In this case the measured TM-losses are close to the calculated TM-losses of slab waveguides. These results indicate that besides depolarization /7/ a weak guidance with respect to depth reduces the TM-losses (compare Fig. 3).

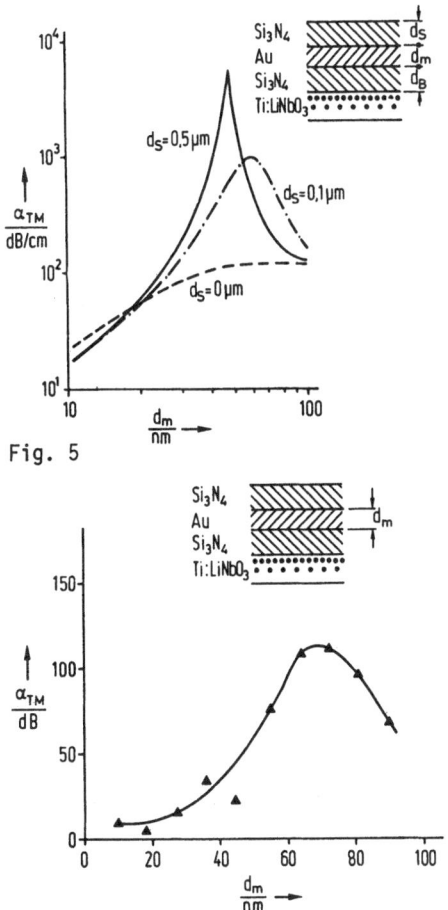

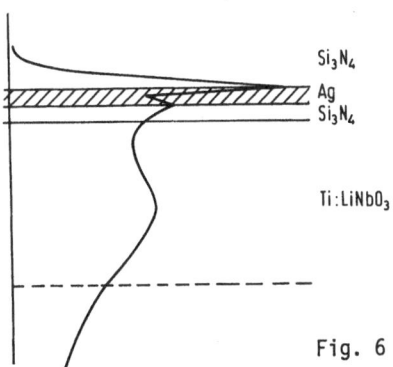

Fig. 5

Fig. 5
Calculated TM-losses of a polarizer with $Si_3N_4/Au/Si_3N_4$ overlay
versus metal thickness d_m. The
diffusion depth of the $Ti:LiNbO_3$
waveguide is $x_D = 1\,\mu m$, the Si_3N_4
buffer layer thickness is $d_B = 50\,nm$

Fig. 6
Calculated field distribution of
a multilayer polarizer with
$Si_3N_4/Au/Si_3N_4$ overlay (slab
waveguide)

Fig. 7
Experimental results of a single-
mode stripe waveguide polarizer
with $Si_3N_4/Au/Si_3N_4$ overlay versus
metal thickness d_m. The Si_3N_4
buffer layer thickness is $d_B = 40\,nm$,
and the thickness of the Si_3N_4
superstrate is $d_s = 240\,nm$

In Fig. 5 theoretical results for a more complicated multilayer polarizer
are represented. A $Si_3N_4/Au/Si_3N_4$ arrangement provides phase-matching by
two additional parameters, the thickness of the gold film d_m and the thickness of the Si_3N_4 superstrate d_s. The maximum TM absorption can be improved
by a factor up to 60, if a $Si_3N_4/Au/Si_3N_4$-system with $d_s = 0.5\,\mu m$ is used.
In this case the thickness of the buffer between waveguide and gold is
$d_B = 50$ nm. Figure 6 shows the calculated field distribution of this
structure. One recognizes the coupling of the slab waveguide mode to a
symmetric plasmon wave in the gold film. This theoretical result is verified by a measurement represented in Fig. 7. The thickness of the buffer
layer is $d_B = 40$ nm (see the arrow in Fig. 4, d_s is chosen to 240 nm. The
thickness of the gold layer is varied between 10 nm and 90 nm. The TM-losses
reach a maximum for a certain thickness d_m of the metal layer, for large
values d_m the superstrate has no influence.

4. Conclusions

Single mode stripe waveguide polarizers on $LiNbO_3$ are fabricated with TM-attenuation of more than 100 dB/cm. It is shown that dielectric/metal/ dielectric overlays on $Ti:LiNbO_3$ waveguides yield additional design flexibility.

5. Acknowledgement

This work has been supported by the Deutsche Forschungsgemeinschaft.

6. References

/1/ L.L. Buhl, Electron. Lett. 19 (1983), 659

/2/ M. Papuchon et al., Electron. Lett. 19 (1983), 612

/3/ T. Findakly, B. Chen, Electron. Lett. 20 (1984), 128

/4/ Uehara, Izawa, Nakagome, Appl. Opt. 13 (1974), 1753

/5/ M. Papuchon, S. Vatoux, IOOC '83, Tokyo, 1983, Post-deadline paper

/6/ D.E. Gray, "American Institute of Physics Handbook", McGraw-Hill, New York (1972), section 6

/7/ J. Ctyroky, J. Janta, J. Schröfel, ECOC '84, Stuttgart, Conference Proceedings

Greatly Reduced Losses for Small-Radius Bends in Ti:LiNbO$_3$ Waveguides

S.K. Korotky, E.A.J. Marcatili, J.J. Veselka, and R.H. Bosworth

AT&T Bell Laboratories, Crawfords Corner Road, Holmdel, NJ 07733, USA

We report the demonstration of a new concept that permits the fabrication of low-loss Ti:LiNbO$_3$ waveguide bends with radii much smaller than previously achieved.

1. Introduction

Advances in the quality and capabilities of Ti:LiNbO$_3$ waveguide devices for optical communication, signal processing, and sensing applications are increasing at a very fast pace. Already, small scale device integration has been demonstrated. To take full advantage of the features this technology offers, and to make optimal use of the substrate material, an efficient means of on- and off-chip interconnection of devices is essential. Optical waveguide bends will consequently play an increasingly important role as the complexity of circuits increases. The ability to make small-radius, low-loss bends will not only increase the number of devices that can fit on a chip, but will permit the realization of waveguide structures such as steep angle, low crosstalk waveguide intersections and ring resonator/filters.

Here we report the demonstration of a technique recently proposed by Marcatili to greatly reduce the bending loss for small-radius bends in Ti:LiNbO$_3$ waveguides. An S-bend transition connecting parallel waveguides offset laterally by 100μm and longitudinally by 1.5mm with an excess loss of 0.1dB was achieved at 1.48μm wavelength. The corresponding radius of curvature of 5.5mm is several times smaller than had previously been obtained [1].

2. Device Design and Measurements

Qualitatively, the technique proposed by Marcatili makes use of micro-prisms (triangles) of raised index placed in the waveguide path in the bend region, as shown in Figure 1, to equalize the optical path lengths for the inside and outside edges of the waveguide mode. In addition, a planar region of raised index is situated near the inner edge of the bend at the base of the triangles to compensate for the tendency of the mode to move toward the outside edge of the bent waveguide. We refer to the concept as the *CROWNING* technique (Controlling Radiation from Optical Waveguides by Notching the Index

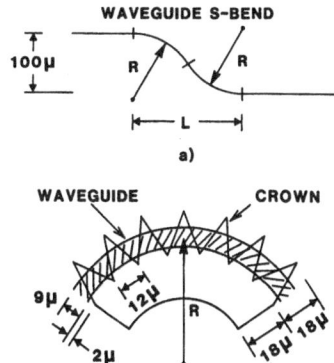

Figure 1   Schematic diagram of waveguide S-bend and application of the CROWNING concept

in the Guide), because of the similarity of the compensating structure to a crown. It is straight-forward to show that the optimum or design radius, R_D, is proportional to the height of the triangles, D, and the bulk index, n_b, and inversely proportional to the index increase, Δn, provided by the triangles, i.e. $R_D = \dfrac{Dn_b}{\Delta n}$. The effect of the *CROWN* structure may also be analyzed by considering the effective radial index distribution of the bent guide [2]. In this representation, the barrier in the index distribution through which tunneling occurs must be made as deep and as wide as possible to avoid bend radiation.

To test the proposed *CROWNING* technique, we have chosen the $1.5\mu m$ wavelength region. Just as the case for conventional dielectric waveguide bends [1], we expect the loss for bends employing the *CROWNING* concept to be a sensitive function of the confinement of the single-mode waveguide. Consequently, we designed [3] the waveguide diffusion conditions for this experiment such that at $1.56\mu m$ wavelength the single-mode waveguides would be close to the cutoff for the second order mode. The diffusion conditions chosen are also compatible with low-loss, low-voltage active devices. For example, using the same parameters, we have fabricated [4] a single coupling length (15mm) directional coupler switch on a 3cm long lithium niobate crystal with a total fiber/device/fiber insertion loss of 3.2dB (of which approximately 3dB is coupling loss) and a switching voltage of 7V.

To implement the *CROWNING* technique, the crown-like pattern and the optical waveguides were fabricated by the diffusion of titanium into the lithium niobate crystal in a double-diffusion process. First, the crown structures were patterned and diffused. Next the waveguides were patterned over the crowns and the crystal was baked again. In this manner we ensured that the crowns penetrated more deeply into the crystal than did the waveguide; the crowns thereby more completely encompassed the optical mode profile in depth. Both diffusions were carried out at $1025^{\circ}C$ for 6hr.

The initial waveguide strip width and titanium thickness were $9\mu m$ and 950Å. We chose both the heights of the triangles, D, and the width of the planar region to be twice the waveguide strip width, i.e. $18\mu m$. The bases of the triangles were $12\mu m$ in length and were positioned $2\mu m$ from the inner edge of the waveguide bends. This is illustrated schematically in Figure 1. The titanium thickness used for the crown was 1000Å. Based on the same model used to design the waveguides, the index change of the crown region near the crystal surface is estimated to be $\Delta n \cong 1 \times 10^{-2}$. We therefore deduce from the design rule that the design radius is $R_D = 4mm$. In Figure 2 we show a photograph of the Ti:LiNbO$_3$ waveguide bend and crown after the diffusions.

Figure 2 Photomicrograph of Ti:LiNbO$_3$ waveguides with CROWN after diffusions

The experiment to verify the *CROWNING* proposal consisted of measuring the losses of the TM mode for constant radius S-bends, with and without the crown, and also for straight reference waveguides. The structures were fabricated on a single z-cut, y-propagating lithium niobate crystal. The lateral offset of the parallel input and output guides was fixed at $100\mu m$ and the longitudinal transition length was varied between 0.5mm and 4.0mm, corresponding to radii between 0.65mm and 40.0mm . For all the S-bends, however, the crowns corresponding to the fixed design radius, $R_D = 4mm$, were used. The reason for designing the experiment in this fashion was to provide a clear signature of the effect. For example, for S-bend radii $R >> R_D$, we expect the crown to be detrimental because of mode mismatch [2]. Near $R = R_D$ there should be a reduction in loss if the crown works. Finally, for

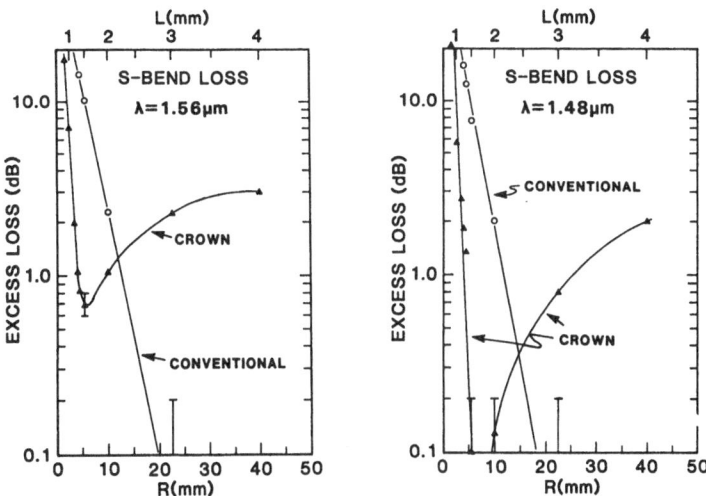

Figure 3   Measured excess loss with and without CROWNING technique

$R < < R_D$ the crown is expected to be less effective, and the loss resulting from bend radiation should approach that of the conventional S-bend.

In Figure 3a we plot the measured excess loss incurred for traversing the bends as compared to a straight waveguide at $1.56\mu m$ wavelength. Evidence that the waveguides are relatively strongly confining at this wavelength is provided by the result that the conventional S-bend loss is lower than 0.1dB for a 20mm bend radius. We observe that near the design radius, the excess loss has been drastically reduced using the crowns from greater than 10dB to less than 0.7dB . The radius of minimum excess loss using the crowns occurs at $R=5.5mm$.

Knowing the importance of waveguide confinement in controlling bend loss, we repeated the measurements on the same device crystal at $1.48\mu m$ wavelength. The waveguides were observed to be single-mode at this wavelength, but were multimode at $1.32\mu m$ wavelength. The results for these measurements are presented in Figure 3b. We find a further reduction of the excess loss near the design radius at this wavelength. At $R=5.5mm$ the loss was measured to be 0.1dB $\pm$ 0.1dB .

3. Summary

In conclusion, we have verified a new concept for permitting low-loss, small-radius Ti:LiNbO$_3$ waveguide bends. An S-bend transition connecting parallel waveguides offset by $100\mu m$ laterally and 1.5mm longitudinally was achieved with an excess loss of 0.1dB at $1.48\mu m$ wavelength.

REFERENCES

1.   W.J. Minford, S.K. Korotky, and R.C. Alferness, "Low-loss Ti:LiNbO$_3$ waveguide bends at $1.3\mu m$," IEEE J. Quantum Electron., *QE-18*, pp. 1802-1806, 1982.

2.   E.A.J. Marcatili and S.K. Korotky, "Sharp bends and fast tapers for integrated optics," to be published.

3.   S.K. Korotky, W.J. Minford, L.L. Buhl, M.D. Divino, and R.C. Alferness, "Mode size and method for estimating the propagation constant of single-mode Ti:LiNbO$_3$ strip waveguides," IEEE J. Quantum Elecron., *QE-18*, pp. 1796-1801, 1982.

4.   J.J. Veselka and S.K. Korotky, "Characteristics of Ti:LiNbO$_3$ waveguides for $1.56\mu m$ wavelength," to be published.

Reduction of Bend-Losses in Integrated Optics Devices

W. Döldissen, H. Heidrich, and D.. Hoffmann

Heinrich-Hertz-Institut für Nachrichtentechnik Berlin GmbH, Einsteinufer 37
D-1000 Berlin 10, Germany

L. Thylén and B. Lagerström

Fiber Optics and Line Transmission, Telefon AB L M ERICSSON,
S-12625 Stockholm, Sweden

Losses at curved/straight waveguide transitions are investigated experimentally and by BPM simulations by introducing
lateral offsets compensating for mode displacement. A loss
reduction up to 2 dB is obtained.

Introduction

In order to use the chip area of integrated optics devices
efficiently, waveguide bends with small radii are indispensable
for interconnecting different functional elements. In material
systems, such as $Ti:LiNbO_3$, where only small index steps are
obtainable, different schemes for reducing the (equivalent) bend
radius have been suggested, such as "directional coupler waveguide offsets" /1/, coherent coupling /2/, different distributions of curvature and curvature changes (circular arcs,
$\sin(x)-x$, $\sin(x)$, etc.). Such schemes are generally possible
to transfer to semiconductor waveguides, and the larger index
steps obtainable make it, in principle possible to reduce the
radius of curvature in comparison to $Ti:LiNbO_3$ waveguides, which
is important in view of the limited wafer size for semiconductors.

As is well known /3/, the losses incurred in a single mode waveguide exhibiting directional changes can be split into two parts:
one originating from changes of the radius of curvature (transition losses) and one due to the curvature itself (pure bend
loss). The transition loss is associated with the displacement
of the modes at the boundary between two differently curved
sections.

In this paper we apply a method described in /5/ for reducing
transition losses in optical waveguides by lateral waveguide offsets.

Reduction of transition losses

We investigate the structure shown in fig. 1 representing a
double circular bend with radius R which accomplishes a lateral displacement x over a length $L = (4Rx-x^2)^{1/2}$. With the in-
clusion of offsets (c and 2c) the transition losses are reduced
by improved mode matching (but not entirely eliminated due to
asymmetry of the guided fields and leaky modes) such that almost
only the pure bend loss remains. For a transition between a
straight and a curved waveguide, the loss is given by the following equation /5/:

$$L_t/dB = 10 \text{ lg } e \quad (d/w_0)^2 \tag{1}$$

where $d = \dfrac{1}{R} (N w_0^2 / \lambda_0)^2$ is the beam displacement,

N the effective index, w_0 the gaussian spot size radius,
λ_0 the free space wavelength, R the radius of curvature of
the bent guide.

According to this equation, the loss reduction for an S-bend in
a Ti:LiNbO$_3$ waveguide (w_0 = 4 µm, N = 2.217, λ_0 = 1.3 µm) can
be estimated to exceed 1 dB for bend radii R⪆ 9 mm. At
smaller radii still pure bend loss prevails.

For a bent step index slab waveguide, on condition that the
straight guide fundamental mode is still a good description of
the bent guide fundamental mode, the following equation taken
from /6/ holds for the loss per unit length:

$$L_B = (\gamma \kappa)^2 / [\beta (1 + \gamma d) (n_1^2 - n_2^2) k_0^2] \exp (2 \gamma d)$$

$$\exp -(2/3)(\gamma^3 / \beta^2) R \quad 10 \log e \tag{2}$$

where 2 d is the slab width, n_1, n_2, N are the core, cladding,
and effective indices, R is the bend radius, $k_0 = 2\pi / \lambda_0$,
$\gamma = k_0 (N^2 - n_2^2)^{1/2}$, $\kappa = k_0 (n_1^2 - N^2)^{1/2}$, $\beta = N/k_0$.

Since equ. (1) is valid for gaussian fields and equ. (2) per-
tains to step index guides, and these equations generally in-
volve a number of approximations, a useful alternative applicable
to the graded index Ti:LiNbO$_3$ waveguides is the use of the beam
propagation method, together with the effective index method /7/.
This circumvents the problem associated with a detailed optimization based on the proper field solutions, treats the transition
and pure bend losses in a unified fashion, incorporates any re-
coupling of power to the bent guides, and also gives information
on crosstalk. Below, this formalism is used for deriving some
theoretical results.

Experimental Results

For this study we fabricated groups of waveguides displaced by
x = 50 µm via circular S-shaped bends (cf. fig. 1). The bend
radii R varied between 5 and 30 mm and, for each radius, the offsets at transitions between segments were changed in the range
of c = 0 ... 3 µm in steps of 0.3 µm. The waveguide width was
5 µm. An evaporated Ti-film (thickness 58 nm) was patterned by
photolithography using an e-beam written mask. The Ti indiffusion was carried out at 1025°C during 5 h. Light (λ_0 = 1,3 µm,
TM-polarized was coupled in and out at the polished endfaces
of the chip by microscope objectives. Output power was measured
with a Ge-photodiode. For evaluation, the average of the attenuation data, taken for three different samples, were referenced
against a straight waveguide on the same chip. The excess loss
per unit length due to the bends is shown in fig. 2 for zero and
optimum offset. For the latter case, the contribution of tran-

211

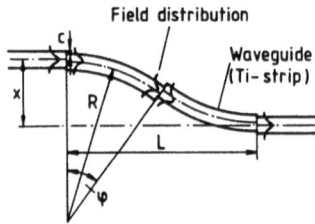

Fig. 1: S-shaped waveguide
transition with offsets
(cf. text)

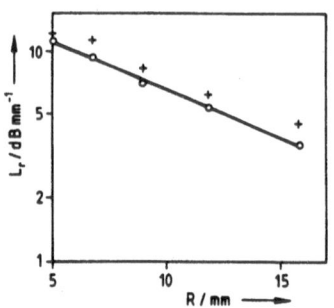

Fig. 2: Bend loss without (x)
and with (o) optimum offset
vs. bend radius

sition loss (equ. (1)) is eliminated and the data represent the
pure bend loss, for which we observe an exponential dependency
on the bend radius in agreement with equ. (2).

The measured loss reductions due to optimized transition matching range from 0.25 dB to 2 dB (for the entire S-bend). The
data for loss reductions are compiled in table 1 as well as
BPM results. Fig. 3 shows a plot of the measured transition
loss data,together with the results of the BPM-simulations for
an index profile giving approximately the same spot size as that
measured. A reasonable agreement with the R^{-2}-dependency as inferred from /5/ is found, however, as explained in /8/ and also
shown by our BPM simulations, this dependency cannot be expected
for very short bend radii. In this regime of small bend radii,
the calculations in /8/ show the attenuation to level off, an
effect that is also observed in our experiment. The discrepancy
from the R^{-2} behaviour may be attributed to index and field
asymmetries at small radii (R ≲ 5 mm) and it should be borne in
mind that equs. (1), (2) are valid only for large radii anyway.

A representation of the loss improvement as a function of the
applied offset (fig. 4) shows a pronounced maximum,indicating
the displacement of the fundamental mode as discussed in /4/.

Table 1: Experimental and calculated loss
         reductions

R/mm	optimized offset/μm	measured loss reduc-tion/dB	calculated loss reduc-tion with BPM/dB
5	0.9	0.36	–
6.7	2.7	2.02	1.08
8.9	2.1	1.13	0.93
11.9	2.1	0.88	0.62
15.8	1.8	0.55	0.41
28.1	1.2	0.25	0.10

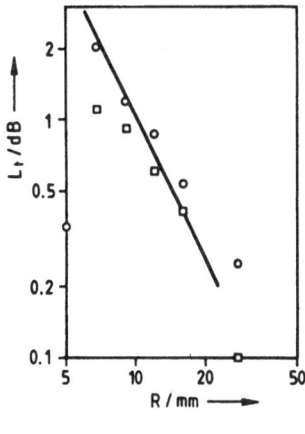

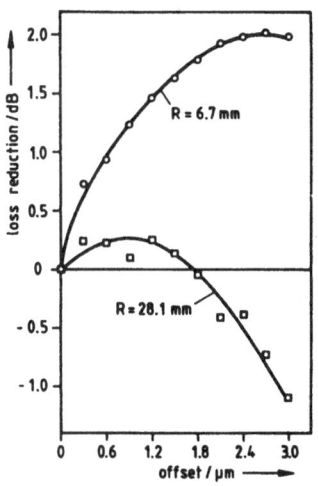

Fig. 3: Transition loss
vs. bend radius measured
(o) and calculated (□)

Fig. 4: Measured loss reductions vs. guide offset for
two different radii

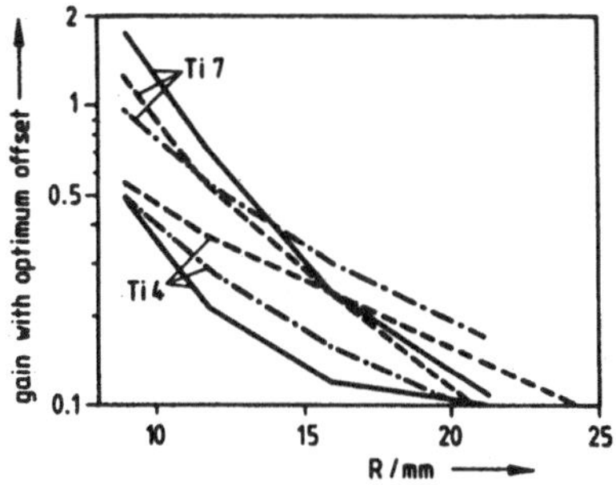

Fig. 5: BPM simulations for two different index profiles
(Ti-4 and Ti-7, cf. text). Dashed lines: achievable gain
at optimum guide offset. Solid lines: lowest bend loss at
optimum offset computed for total transition length
L=1500 μm with guide offset only between the two curved
sections. Dashed-dotted lines: values calculated after /5/.

Discussion and Conclusions

The above experimental data imply the possibility to significantly reduce the bend losses at small radii (R $\lesssim$ 10 mm). Unfortunately, for these small radii, large pure bend losses are observed. In order to see the potential of this method for reducing losses, we plot in fig. 5 BPM simulations for two profiles,
Ti-4 and Ti-7, with $N-n_s$ = 0.0037, w_o = 3.82 μm, and $N-n_s$ =
0.0031, w_o = 4.26 μm, respectively.

It is obvious from fig. 5, that for the Ti-4 profile at R = 9 mm
a reduction of the 1 dB loss obtained for Ti-7 to a residual
pure bend loss of 0.25 dB is possible,provided that offsets are
also implemented between the straight and curved sections. This
clearly shows the merit of the offset method for judiciously designed waveguides.

The Ti-4 profile was used to calculate the transition loss data
of fig. 3. It is, however, found that this profile does not give

as large a pure bend loss as do the actual experimental profiles. The spot sizes (for Ti-4 and the experimental waveguides) are very similar, and this is sufficient for the BPM formalism to give good agreement, as shown in fig. 3. However, a detailed optimization relies on an accurate description of the index profile itself.

Comparisons between theory and experiment using the experimentally determined profile will be reported, as will investigations relating to the total optimization of S-bends and waveguides.

References

1 H.A. Haus, C.G. Fonstad: "Three Waveguide Couplers for Improved Sampling and Filtering", IEEE J. Quantum Electron., QE-17 (1981), pp. 2321-2325

2 L.M. Johnson, D. Jap: "Theoretical Analysis of Coherent Coupled Optical Waveguide Bends", Appl. Opt. 17 (1984), pp. 2988-2990

3 W.A. Gambling, h. Matsumura, C.M. Ragdale: "Curvature and Microbending Losses in Single Mode Optical Fibres", Opt. Quant. Elec. 11 (1979), pp. 43-59

4 D. Marcuss: "Field Deformation and Loss Caused by Curvature at Optical Fibres", J. Opt. Soc. Am. 66 (1976), pp. 311-320

5 E.G. Neumann: "Curved Dielectric Optical Waveguides with Reduced Transition Losses", Proc. IEE-H, 129 (1982), pp. 278-280

6 D. Marcuse, in Light Transmission Optics (Van Nostrand Reinhold Company, New York, 1972, Chapter 9)

7 L. Thylén, B. Lagerström, P. Svensson, A. Djupsjöbacka, G. Arvidsson: "Computer Analysis and Design of Ti:LiNbO$_3$, Integrated Optics Devices and Comparison with Experiments", Proc. ECOC 83, Elsevier Amsterdam (1983), pp. 425-428

8 R. Beats, P.E. Lagasse: "Loss Calculation and Design of Arbitrarily Curved Integrated-Optic Waveguides", J. Opt. Soc. Am. 73 (1983), pp. 177-182

Integrated Optical Wavelength Multiplexer/Demultiplexer for Optical Communication

H.-P. Nolting, D. Hoffmann, and M. Schlichting

Heinrich-Hertz-Institut für Nachrichtentechnik Berlin GmbH, Einsteinufer 37
D-1000 Berlin 10, Fed. Rep. of Germany

We propose a new multiplexer/demultiplexer structure with
tunable center wavelength and a bandwidth between several nm
to 100 nm on InP by varying waveguide dimensions.

Introduction

In optical communication, wavelength multiplexers/demultiplexers
(MULDEX) which are compatible with the technologies of integrated optics (IO) are required. Previous concepts for the realization of such filters are:

- type A: asymmetric directional coupler with equal effective
  indices at the center wavelength realized in LiNbO$_3$ /1/ and
  on InP /2/

- type B: TE— TM mode converter/wavelength filter ($\Delta\mathcal{H}$-coupler)
  realized in LiNbO$_3$ /3/

- type C: asymmetric directional coupler with contradirectio-
  nal coupling by a bragg-grating /4/

- type D: TE — TM mode converter realized in GaAs/GaAlAs (not
  designed as a filter) /5/.

We will compare them with the proposed filter (type E) in
table 1 and furthermore will give the design-rules for the new
structure and some calculated examples for the quaternary material system InGaAsP.

Design and theoretical description of the MULDEX

We propose a new MULDEX which is composed of two nonidentical
optical waveguides with a periodic structure of N sections, with
the length Λ made up of two subsections with the coupling co-
efficient $\mathcal{H} = \mathcal{H}_0$ and $\mathcal{H} = 0$ along the lengths $l_\mathcal{H}$ and l_0, respectively (Fig. 1). The decoupling can be achieved, for in-
stance, by etching grooves between the adjacent waveguides. In
analogy to the $\Delta\beta$-reversal directional coupler /6/ and to the
TE— TM mode converter/filter with a $\Delta\mathcal{H}$-reversal-structure /1/
we call the device $\Delta\mathcal{H}/2$-coupler. We assume two modes travelling
along a waveguide structure with different propagation constants $\beta_A \neq \beta_B$ and different mode dispersion

$$\frac{\partial \beta_A}{\partial \lambda} \neq \frac{\partial \beta_B}{\partial \lambda}$$

215

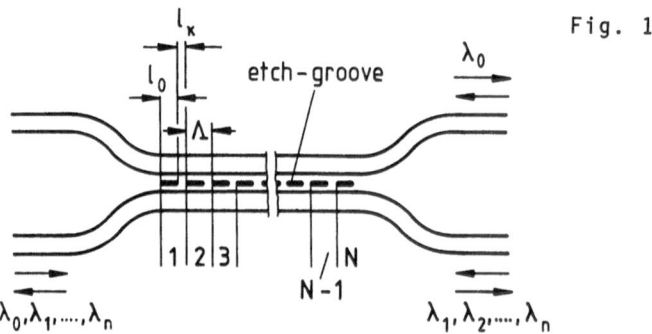

Fig. 1

Δκ/2 - coupler as Mux / Demux

Let the phase-mismatch between the two modes for isolated wave-guides be

$$\delta = \frac{1}{2}\left(\beta_A(\lambda) - \beta_B(\lambda)\right)$$

at the wavelength λ and the ratio of δ and $\varkappa$ be $r \gg 1$. From coupled mode theory /7/ it is well known that such a strong asymmetric coupler with constant coupling coefficient along the whole structure has a periodic energy exchange with a maximum of $I^{\oplus} = 1/(1+r^2)$. By introducing a phase matching periodic structure, however, it is possible to couple the energy at the wavelength λ completely from mode A to B if the correct section lengths $l_{\varkappa}$, l_0 and total device length L_c are chosen. We calculate the crossover intensity $I^{\oplus}(N,\delta,\varkappa, l_0^c, l_{\varkappa})$ of the MULDEX by a matrix algorithmus in a similar way as described in /6/. The result is

$$I^{\oplus} = \frac{1}{1+r^2} \sin^2 \psi_1 \frac{\sin N\theta}{\sin \theta}$$

$$\theta = \arccos\left\{ \cos\left(\psi_1 - \psi_2\right) - \left(1 + \frac{r}{\sqrt{1+r^2}} \cdot \sin \psi_1 \cdot \sin \psi_2\right)\right\}$$

$$\psi_1 = l_{\varkappa} \sqrt{\varkappa^2 + \delta^2} \qquad \psi_2 = l_0 \cdot \delta$$

The design rules for the MULDEX can be derived by analysing the condition for complete crossover at λ_0: $I^{\oplus}(\delta_0, \varkappa_0) = 1$ with $\delta_0 = \delta(\lambda_0)$ and $\varkappa_0 = \varkappa(\lambda_0)$.

The lengths of the two subsections are

$$l_0 = \frac{\widetilde{\pi}}{2\,\delta_0} \qquad\qquad l_{\varkappa} = \frac{\widetilde{\pi}}{2\sqrt{\delta_0^2 + \varkappa_0^2}} \approx l_0$$

They result from the phasematching conditions $\psi_1 = \psi_2 = \widetilde{\pi}/2$. This condition means that in the coupling section the maximum energy-transfer is obtained, and the reverse-coupling is suppressed in the zero-coupling section, meanwhile a phase-shift of $\widetilde{\pi}$ occurs between both modes. Hence, in the next section the phase is adjusted such that energy-transfer takes place again in the same direction as in the preceding coupling section.

The resulting section number N for complete energy-transfer, or equivalently the device-length L_c are:

$$N = INT\left(\frac{\tilde{\pi}}{2} \frac{\delta_0}{\varkappa_0}\right) \qquad \qquad L_c = \tilde{\pi}^2/2\varkappa_0$$

The device length depends only on the coupling constant $\varkappa_0$.
For this MULDEX a bandwidth of

$$\Delta \delta_{FWHM} = \frac{\delta_0}{N} = \frac{2}{\pi}\varkappa_0$$

results. From this condition finally, the bandwidth with respect
to the wavelength can be calculated if the dispersion of δ is
known:

$$\Delta \lambda_{FWHM} = \frac{\delta_0}{N}\left(\frac{\partial \delta}{\partial \lambda}\right)_{\lambda_0}^{-1}$$

It can be shown that it is not necessary to realize the value
$\varkappa_0 = 0$ in one part. It is sufficient that the coupling coefficients differ:

$$\varkappa_1 = v \cdot \varkappa_2 \quad \text{with} \quad 0 \leq v < 1$$

however, at the cost of increased device length (by a factor
of $1/1-v$).

Generally it is possible to modify the passband characteristic
with respect to the side lobe and peak intensity and the bandwidth by chirping the periodic structure (see e.g. /4/).

For tuning the center wavelength λ_0 by the electrooptic effect one can use control electrodes on each waveguide to shift
the phase mismatch δ in both directions by an amount of

$$\Delta \delta_{EO} = \frac{1}{2}U \cdot \Delta \phi_N$$

where U is the applied voltage on one of the waveguides and $\Delta \phi_N$
is the phase-shift per unit length and unit voltage. $\Delta \phi_N$
can be calculated by common methods for quaternary rib waveguides in InGaAsP/InP /8/. Because of the mode dispersion, the
center wavelengths for TM and TE waves are slightly different.
However, it is possible to equalize the center wavelengths with
the electrooptic effect, which influences only the TE mode for
a usual crystal orientation on III-V-compounds.

Application to the quaternary material system InGaAsP

In order to illustrate the described theory, we calculate as an
example the performance of a MULDEX in the material system
InGaAsP/InP. Fig. 2 shows the cross-section of a possible
coupler structure to be fabricated from only one single quaternary layer (bandgap equivalent to $\lambda_g = 0.97$ µm) of the thickness t_F on an InP substrate. For the numerical calculation of
the waveguide parameters, we use the formula for the refractive indices of the q-layers given by /9/. The phase constants
of the two decoupled waveguides can be calculated by using the
ordinary wave equations /10/. They lead to the phase mismatch
parameter δ. For the coupling constant $\varkappa_0$ we use the approach

$$\varkappa_0 = \sqrt{\varkappa_A \cdot \varkappa_B}$$

where $\varkappa_A$ and $\varkappa_B$ are the coupling constants of a symmetric coupler with the same center distance but build up of two equal

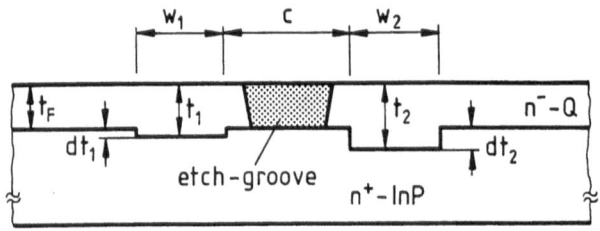

Δκ/2 - coupler for InGaAsP/InP

Fig. 2   Cross section

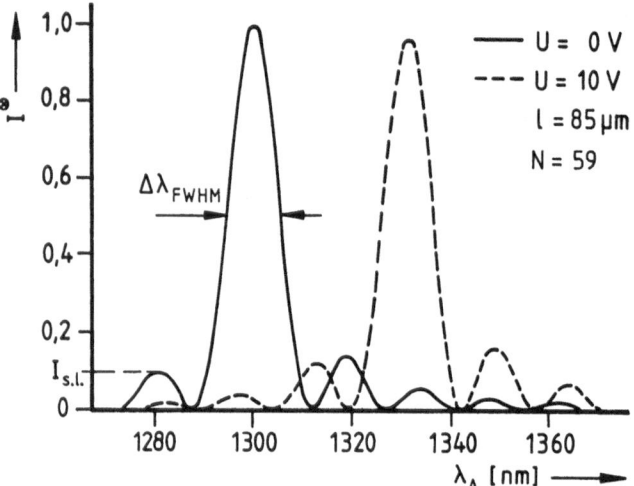

Passband of passive and electrooptically tuned filter   Fig. 3

waveguides A and B /10/. In our calculations we assume that the
two embedded rib waveguides have equal widths $w_2 = w_2 = 3$ μm
and a separation of c = 3 μm and that the first waveguide has
a rib height of $dt_1 = 0.1$ μm. By variation of the film thickness t_F and the rib height of the second waveguide, the bandwidth at $\lambda = 1.3$ μm ranges from a few nm to more than 100 nm
as can be seen in fig. 4. In fig. 5 the corresponding lengths
of the filters are displayed. For a special example ($t_F = 0.9$ μm,
$dt_2 = 0.5$ μm, $l_0 = l_{\varkappa} = 85$ μm and N = 59) we show in fig. 3
the passband characteristic for TE with a bandwidth $\Delta\lambda_{FWHM} =$
14 nm, a side lobe intensity $I_{SL} = 0.14$ and a tunability
of $\Delta\lambda_{tune} = 3\Delta\lambda_{FWHM}$ by 10 V using $\Delta\phi_N = 0.14$ rad/Vmm /11/.
Polarization independence can be achieved by applying 20 V to
the device.

Comparison

In comparison with the concepts mentioned in the introduction
and listed in table 1 the MULDEX has the advantages to be more

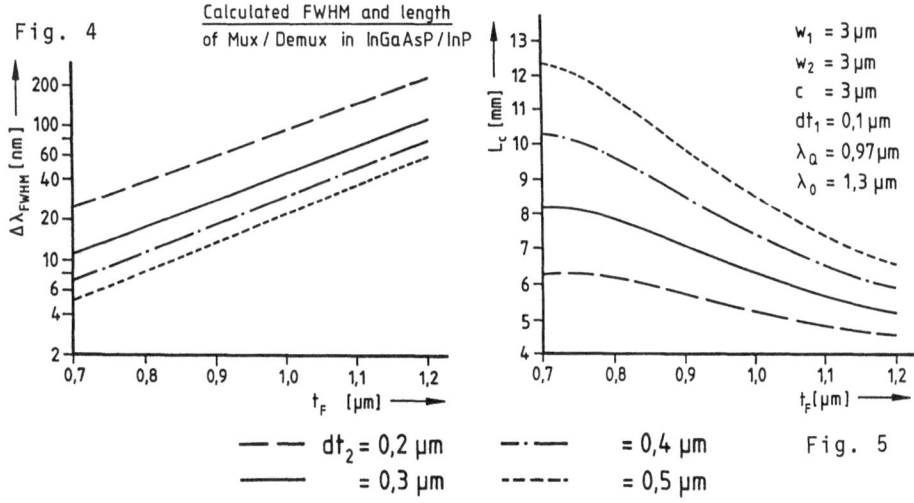

Fig. 4

Calculated FWHM and length
of Mux / Demux in InGaAsP/InP

$w_1 = 3\,\mu m$
$w_2 = 3\,\mu m$
$c = 3\,\mu m$
$dt_1 = 0,1\,\mu m$
$\lambda_Q = 0,97\,\mu m$
$\lambda_0 = 1,3\,\mu m$

Fig. 5

— — $dt_2 = 0,2\,\mu m$ —·— $= 0,4\,\mu m$

——— $= 0,3\,\mu m$ ----- $= 0,5\,\mu m$

Table 1.

Type	Principle	Material	Bandwidth	Tunable Param.	Ref.
A	codirec. coupler	$LiNbO_3$ InGaAsP/InP	200 nm 2 nm	λ_0 λ_0	/1/ /2/
B	TE/TM-mode conv. ($\Delta\kappa$)	$LiNbO_3$	7 nm	$I^{\otimes}$	/3/
C	contradir. coupler		$\approx 0,2$ nm		/4/
D	TE/TM-mode conv. ($\Delta\kappa$ /2)	GaAs/GaAlAs	200 nm	$I^{\otimes}$	/5/
E	codirec. coupler ($\Delta\kappa$ /2)	[InGaAsP/InP] calculated	2..200 nm variation by design	λ_0	this paper

flexible and easier to fabricate. Type A needs two materials
to realize narrow filter characteristics in quaternary materials; type C has a periodical structure in the submicron range
with the difficulty to achieve the center wavelengths precisely. Generally, type B has to be completed to a MULDEX by additional components (e.g. mode splitter) and furthermore always
needs a biasing voltage.

Conclusion

A new MULDEX structure has been proposed that could be fabricated from a single quaternary layer on InP. The bandwidth can
be varied between a few and 100 nm by changing the waveguide
dimensions. The center wavelength is tunable by the electrooptic
effect for TE polarization, and thus the device can be made
polarization independent. Since the device is passive, the structure can also be transferred to other materials e.g. glass or
$LiNbO_3$.

Acknowledgment

The work was supported by the Federal Ministry of Research and Technology (BMFT) and the Senate of Berlin (West).

References

1   R.C. Alferness, R.V. Schmidt: Appl. Phys. Lett. 33(2), 161 (1978)

2   S. Lindgren, B. Broberg, M. Öberg, H. Jiang: Int. Conf. on Thin Solid Films, Stockholm, Aug. 1984

3   R.C. Alferness: Appl. Phys. Lett. 36(7), 513 (1980)

4   P. Yeh, H.F. Taylor: Appl. Optics, 19, 1848 (1980)

5   F.K. Reinhart, R.A. Logan, W.R. Sinclair: QE 18, 763 (1982)

6   H. Kogelnik, R.V. Schmidt: QE 12, 396 (1976)

7   A. Yariv: QE 9, 919 (1973)

8   P. Albrecht, H.G. Bach, C. Bornholdt, W. Döldissen, D. Franke, N. Grote, J. Krauser, U. Niggebrügge, H.P. Nolting, M. Schlak, I. Tiedke, R.A. Logan, F.K. Reinhart: 2nd European Conference on Integrated Optics, Florenz, Okt. 1983

9   Y. Suematsu, K. Utaka, K.I. Kobayashi: QE 17, 651 (1981)

10 J.C. Shelton, F.K. Reinhart, R.A. Logan: J. Appl. Phys. 50, 6675 (1979)

11 C. Bornholdt, W. Döldissen, D. Franke, J. Krauser, U. Niggebrügge, H.-P. Nolting, F. Schmitt: to be published on same conference

220

Design of an Integrated Optic Grating Demultiplexer

B. Denturck and P.E. Lagasse

Laboratory of Electromagnetism and Acoustics, University of Gent
Sint-Pietersnieuwstraat 41, B-9000 Gent, Belgium

The design of a monomode demultiplexer, based on a grating positioned in a planar integrated waveguide Y-junction, is described. Measurement results on a demultiplexer fabricated by ion exchange in glass are given.

§ 1. Introduction

The advantages of wavelength multiplexing in optical fibre communication systems are well known and do not need to be discussed any further in this paper. The real problem lies in the succesful implementation of this technique. This requires obviously that wavelength demultiplexers must be made with low insertion loss and good channel separation. This has turned out to be quite difficult,and so many different designs have been reported that it is impossible to list them here. Most of those demultiplexers can be divided into four categories,depending on whether they are intended for monomode or multimode fibres,and whether they are based on planar integrated optic or micro-optic fabrication techniques. In this paper we will restrict ourselves to the problems and trade-offs related to the analysis and design of a planar integrated optic wavelength demultiplexer for monomode systems. As wavelength selective device,a grating positioned in a waveguide Y junction was chosen. The general design considerations of this structure are discussed in more detail in the next paragraph. The calculation of the gratings by means of a special coupled beam propagation method [1],[2], is also described. Finally,fabrication details and measurement results of test structures and devices are reported.

§ 2. General design considerations

The starting point for the design of the grating demultiplexer was the requirement to achieve less than 50nm separation between channels, starting from monomode guide and using a planar integrated optic grating etched in glass. In order to be compatible with standard micro-fabrication techniques, it was decided to define the gratings by contact printing of electron beam masks instead of by holographic recording or by direct electron beam writing. Although this makes fabrication cheaper and simpler, it puts a limit on the smallest periodicity for the grating that can be achieved. A 2µm period, i.e. 1µm line width was chosen. This means that the deflection angle is rather small, which increases the bandwidth of the grating.

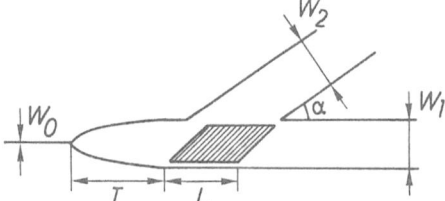

Fig. 1 Lay out of the demultiplexer

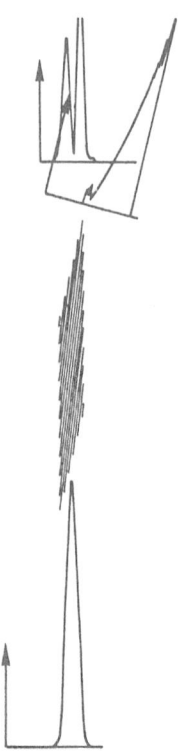

Fig. 2 Transmitted and reflected beam
in the grating

The complete demultiplexer structure is sketched in fig. 1.
It consists of a monomode input guide of width w_0 tapering
over a length T into a much wider waveguide of width w_1. In
this wide waveguide a Y junction is formed by a branching waveguide of width w_2. As schematically shown in fig. 1 a grating,
etched to a depth d, is situated in the Y junction so as to
deflect the light of the selected wavelength into the branch.
The design problem consists in the choice of the different
dimensions mentioned above. Let us now discuss the trade-offs
that need to be taken into account. The coupled beam propagation method [1], [2] was used to calculate the effect of the
length L , the width w_1 and the depth d of the grating on the
bandwidth and on the coupling efficiency. For those calculations the fundamental mode of the waveguide with width w_1 was
used as input beam. Inside the Y junction, this field is[1]
no longer laterally guided to the side of the branch since the
branching angle is larger than the natural diffraction angle
of a beam with width w_1. To the other side, the beam still
sees the edge of the waveguide. In the calculation of the grating, this last guiding effect was neglected. A typical result
of such a calculation is shown in fig. 2 where w_1 = 200µm,
L = 1650µm, α = 14.2° and the coupling coefficient K of the
grating was 1500m^{-1}. For this case 3dB bandwidth of 20nm and
a deflection efficiency of -2.3dB was calculated. It was found
that the 3dB bandwidth varies in an inverse proportion to the

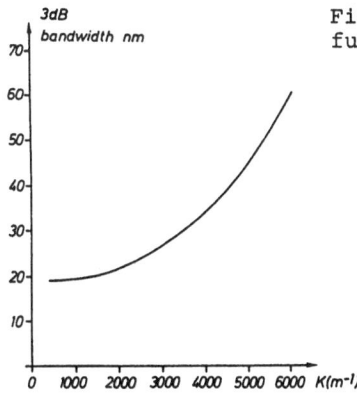

Fig. 3 Calculated 3dB bandwidth as a function of coupling bandwidth

width w_1 of the grating. If we want a 40nm channel separation, w_1 cannot be much smaller than 200µm. If w_1 is reduced, the beam divergence due to lateral diffraction increases, resulting in a decrease of the deflection efficiency. The main disadvantage of a large width w_1 is the greater length of the taper required for an adiabatic transition of the fundamental mode from the monomode guide to the guide of width w_1. This taper was also calculated using the beam propagation method [3]. The coupling constant of the grating cannot be made too large since this increases the bandwidth. In fig. 3 the calculated 3dB bandwidth is plotted as a function of the coupling constant for a grating having the same dimensions as the one described in fig. 2. One can see that this coupling constant should not be made larger than 3000m⁻¹. For a grating length L of 1.65mm this gives a good coupling efficiency, but it is not possible to reduce this length seriously. The width w_2 of the branching guide is 350µm, which seems acceptable for an output guide leading to a detector.

The dimensions discussed above obviously depend on the central wavelength of the demultiplexer. For this experiment 750nm was chosen as the central wavelength. This choice was obviously not dictated by fibre optic systems considerations but by the fact that this is the central wavelength of an oxazine dye laser. Since our primary aim was to test the validity of the design methods and the feasibility of the fabrication methods, the ability to easily and accurately measure the devices was more important than their potential system use. When working at 1.3µm wavelength, and keeping the same 2µm period for the grating, the deflection angle would be increased and all dimensions therefore reduced.

§ 3. Fabrication and measurement results

The device which was finally made had following dimensions : grating period = 2µm; central wavelength = 750nm; w_0 = 4µm; w_1 = 200µm; w_2 = 350µm; length of parabolic taper T^0 = 18mm; α^1 = 14.2°. The structure was defined by means of two electron beam masks. The first one contained the waveguide, the second one the grating. All waveguides were formed by silver ion exchange in glass. The ion exchange consisted of a single

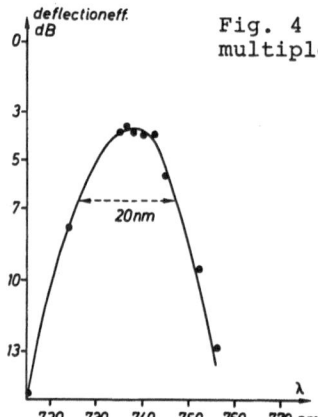

deflection eff. dB

Fig. 4 Measured deflection efficiency of the demultiplexer as a function of wavelength

indiffusion step through an anodized aluminium mask, so as to obtain a maximum of the refractive index at the surface of the substrate. This allows one to obtain a large coupling coefficient for the grating by relatively shallow etching of the surface. The obvious disadvantage is that the waveguides have more loss. The depth to which the grating needs to be etched in order to obtain the given coupling coefficient, can be calculated to a first approximation by means of a coupled wave theory. This approximation is good for shallow gratings, but it has been found that for larger coupling coefficients more exact calculations are needed [4],[5]. Typical etching depths used in the experiments ranged between 30 and 80nm.

Measurements of wavelength selectivity were performed by injecting light at the monomode input waveguide and measuring the light output from the branch waveguide. As can be seen in fig. 4 a 3dB bandwidth of 20nm was obtained and a 10dB bandwidth of approximately 40nm. The deflection efficiency was measured to be about −3dB. Those results are in quite good agreement with the theoretical calculations, and prove that the design method described in this paper can be succesfully implemented.

Acknowledgement

The authors would like to thank W. De Raedt for the fabrication of the e beam masks. B. Denturck would like to thank IWONL for a grant.

[1] J. Van Roey, P.E. Lagasse : "Coupled wave analysis of obliquely incident waves in thin film gratings" in Applied Optics, Vol. 20, p. 423 (1981).
[2] J. Van Roey, P.E. Lagasse : "Coupled beam analysis of integrated optic Bragg reflectors", JOSA, Vol. 72, p. 337 (1982).
[3] R. Baets, P.E. Lagasse : "Calculation of radiation loss in integrated optic tapers and Y-junctions", Applied Optics, Vol. 21, p. 1972 (1982).
[4] J. Van Roey, B. Denturck, P.E. Lagasse : "Guided wave coupling in integrated-optic gratings : normal incidence", IEE proceedings, Vol. 131, Pt. H, p. 282 (1984).
[5] J. Van Roey, Ph.D. thesis, University of Gent, Belgium, 1983.

Integrated-Optical 8×8 Star Coupler in Ti:LiNbO$_3$

W. Döldissen, H. Heidrich, D. Hoffmann, H. Ahlers, A. Döhler, and
M. Klug

Heinrich-Hertz-Institut für Nachrichtentechnik Berlin GmbH, Einsteinufer 37
D-1000 Berlin 10,Germany

The first planar 8x8 star coupler, fabricated on x-cut
Ti:LiNbO$_3$ consisting of twelve interconnected 3-dB-couplers
with almost uniform power distribution is presented.

Single mode star couplers will be one of the key devices in
future optical fibre communication systems. Although planar
waveguide couplers still exhibit higher insertion loss than
fibre star couplers,they offer the advantages of easier eco-
nomic batch fabrication, reproducibility and the potential of
single chip integration, which makes them attractive for implementation in optical communication systems.

We report on the realisation of a planar 8x8 star coupler consisting of fully connected 3 dB-directional couplers /1/, /2/,
/3/.

The device is based on the recently published first integrated-
optical 4x4 star coupler /4/ and extended to eight input and
output ports. Our 8x8 passive, compact integrated optical power
mixing device is fabricated on x-cut LiNbO$_3$ using z-propaga-
tion in order to avoid the materials birefringence for TE and
TM polarised light,and to make the device insensitive to fluctuations of the state of polarisation of the light coupled
from the input fibres.

The remaining small TE/TM dispersion due to the asymmetry of
the waveguide may be reduced by employing a high refractive
index superstrate.

The architecture and the layout of the device is shown in fig.
1 and 2. The waveguides of 6 µm in width are designed for single
mode operation at λ = 1.3 µm. The input and output waveguides
are separated by 180 µm for fibre coupling. The first and second
four subsequent waveguides are connected via four 3 dB-directional couplers forming two 4x4 star couplers /3/. In a second cas-
cade these two 4x4 star couplers are connected with four 2x2
star couplers (3 dB-directional-couplers) forming the 8x8 power
mixing device (Fig. 3). Each path for the light distribution
contains the same number of equivalent bends of radius R = 40 mm
resulting in equal loss performance.

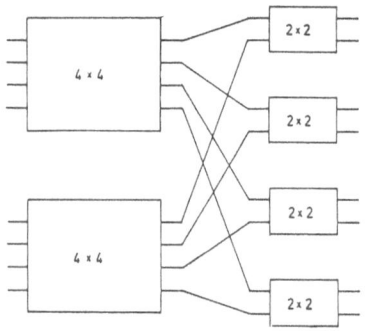

Fig. 1: Architecture of the 8x8 star coupler

Fig. 2: 8x8 star coupler design

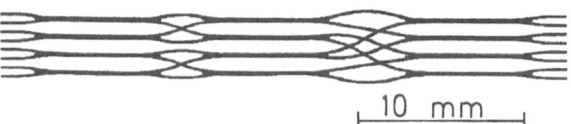

10 mm

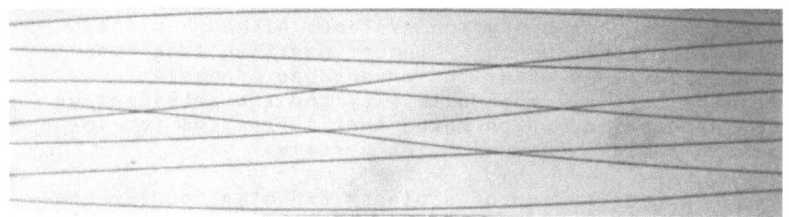

Fig. 3: Photograph of the interesting waveguides between the 4×4 and 2×2 power mixing devices. Waveguide width 6.0 µm, bend radii 40 mm

The waveguide configuration connecting the diametrically oppo-site directional-couplers is shown in fig. 2. The waveguides in-tersect under full angles of about 10° resulting in a theore-tical crosstalk of less then -23 dB.

In order to optimize the 3-dB-coupling at given waveguide sepa-ration of 6.0 µm, we calculated the additional effective inter-action length L_a of the input and output waveguides of the coup-ler

$$L_a \approx \sqrt{\frac{\pi}{2} \frac{R}{\gamma}} = 340 \text{ µm}$$

(R:bend radius; γ :transverse decay constant) and varied the in-teraction length in steps of 50 µm ranging from 1.50 mm to 1.70 mm for the straight section.

The chromium mask was generated by high resolution e-beam-litho-graphy,with a special program for layout providing smooth double-bends and offering device parameter flexibility. The mask pattern was transferred to the Ti-film by contact photolithography using a lift-off-technique.

226

The indiffusion of the 100 nm thick Ti-stripes into $LiNbO_3$ was carried out at 1025° C during 8 h in flowing Ar atmosphere bubbled through water at 95° C, followed by oxidation in flowing O_2 for one hour during cooling. Subsequently a 1 μm thick $LiNbO_3$-superstrate was sputter-deposited in order to reduce the mode asymmetry.

The first device was characterized with respect to polarisation sensitivity and insertion loss. For a selected star coupler figures 4 a) and 4 b) show the intensity distribution at the eight output ports for both polarisaitons. Light was fed into input 1 by a microscope objective,and the nearfield pattern at the output was imaged on an infrared vidicon. Similar results were obtained for the other input ports. The excess loss was approximately 5 dB. Further results will be reported at the conference.

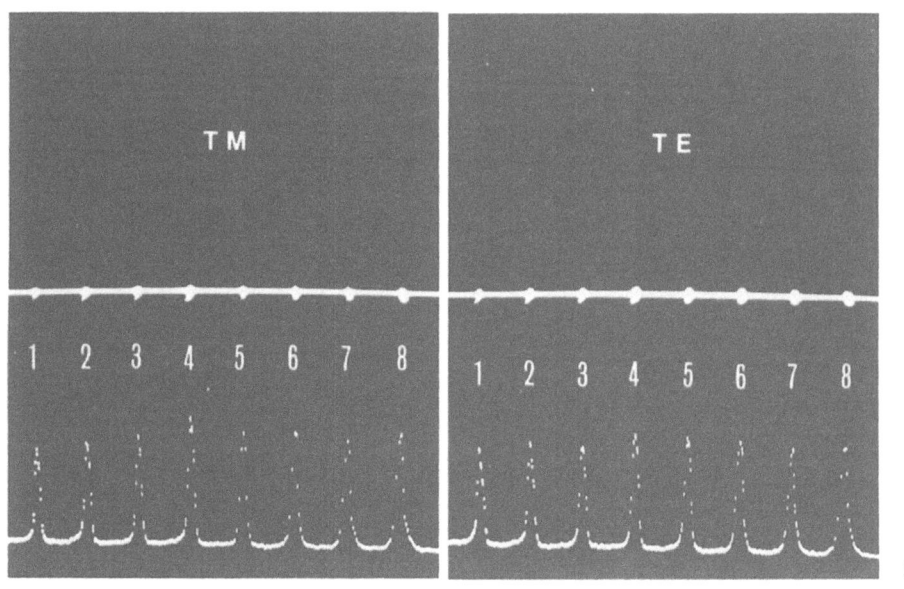

(a)   (b)

Fig. 4: Intensity distribution at the eight outpout ports
        a) for TM- and b) for TE-polarization. Light was
        fed into input 1 of the star coupler. Waveguide
        separation 180 μm

Acknowledgement

The work was supported by the Federal Ministry of Research and Technology (BMFT) and the Senate of Berlin (West).

References

1   Th. Hermes, J. Saniter, F. Schmidt: "Der Aufbau großer mono-
    modaler Sternkoppler", Nachrichtentechn. Z., 37, pp. 636-
    638 (1984)

2   German Patent Application, filed June 21, 1984 (Ser. No.
    P 3423 221.4), Inventors: J. Saniter, F. Schmidt, Th. Hermes

3   M.E. Marhic: "Hierarchic and Combinatorial Star Couplers",
    Opt. Lett., 9, pp. 368-370 (1984)

4   H. Heidrich, D. Hoffmann, W. Döldissen, M. Klug: "Integrated
    Optical 4x4 Star Coupler on $LiNbO_3$", Electr. Lett., 20
    pp. 1058-1059 (1984)

Performances of an Ion Exchanged Star Coupler for Multimode Optical Communications

G. Voirin

Radiall, 101 rue Philibert Hoffmann, F-93116 Rosny-sous-Bois, France

R. Rimet and G. Chartier

Laboratoire de Génie physique, ENSIEG BP 46, F-38402 St. Martin d'Hères, France

We describe a method for fabricating 4-port, 8-port and 16-port couplers for use in multimode fiber communications. Low losses have been achieved.

1. Introduction

Dividing or concentrating optical guided beams are very important operations for optical networks. Several methods have been reported for making the corresponding devices [1], [2], [3]. We have made a component which couples one multimode fiber with N other multimode fibers (N=4, 8, 16). The main problems for such devices are the following : (i) Insertion losses as low as possible, (ii) A good balance between the output intensities, (iii) Easy and cheap mass production. These three points are not obviously compatible. We think that our component represents a good compromise. The method we use is based on a planar technology and ion exchange in glass, and is quite suitable for mass production. In the case of "one to sixteen" ports coupler, we have measured total insertion losses of 4.6 dB at the wavelength of 0.85 micron. The dispersion between the output intensities is as low as $\pm$ 0.5 dB.

2. Fabrication

Ion exchange take place between Ag^+ ions of a molten salt and Na^+ ions of a glass plate. The chemical composition modification,which is a result of this exchange,induces an increase of the index of refraction. In order to make strip-waveguides,we limit the exchange area with a mask etched in an aluminium layer. The guide is obtained by microphotolithography and chemical etching. The component is composed of a succession of Y-shaped waveguides. If the molten salt is just pure silver nitrate, the refraction index increase is about 0.1 which represents to a far too large numerical aperture. To obtain a numerical aperture of 0.2 we use two different methods : (i) The silver nitrate is diluted in sodium nitrate, (ii) A first exchange using pure silver nitrate is followed by a second exchange in pure sodium nitrate. For a better control of the ion exchange and diffusion process, an electric field can be applied ; this has allowed us to bury our waveguides. The application of an electric field is not straightforward, we make use of molten salt electrodes inside of a special PTFE container. Figure 1 shows a typical cross-section of the guide.

3. Devices

The component is composed of a succession of "two-branched" waveguides. When designing the device, we must find a good compromise between the angle of the two branches and the total length. On the side where N waveguides are present, their separation must be at least equal to the cladding diameter (125 microns). To obtain a 16 port divider with an acceptable length (a few centimeters), we choose an angle of 2° between the two branches. A theoretical study, using the well known Beam Propagation Method,

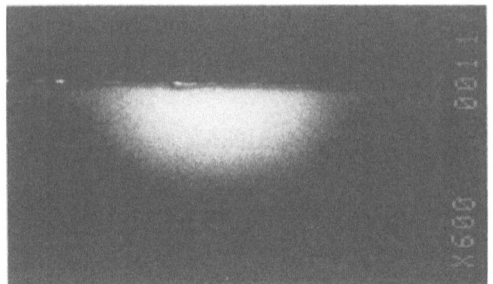

Fig. 1 Electron microscope photograph of a waveguide cross-section

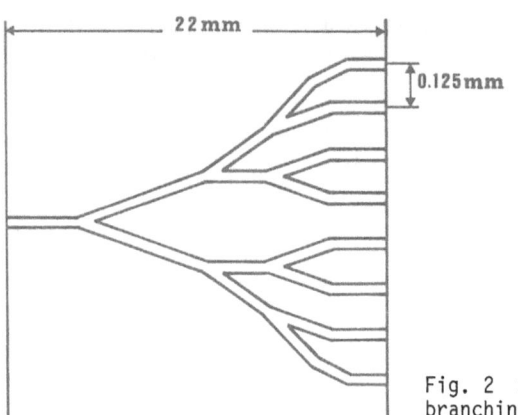

Fig. 2 Schematic pattern of typical branching circuit

has shown that for this angle the losses are 0.2 dB. Figure 2 gives a drawing of a coupler between one fiber and 8 others.

4. Results

To measure the linear losses of our waveguides, we progressively shorten the guide by cutting it several times, and we measure the insertion losses at each step. The measured losses are 0.7 dB/cm.

For testing the couplers, we have measured them in real use. We insert the component between two reels of gradient index multimode fiber having a core diameter of 50 micrometers and numerical aperture of 0.2. The reference is the transmitted power between two butt-joint fibers. The light source is an LED (0.85 micron). We have measured insertion losses of 3dB and 3.5 dB for a 4-port and 8-port divider respectively.

Table 1 lists the ratio between the reference power and each output power of a 16-port circuit. Average ratio is 16.6 dB which gives a total insertion loss of 4.6 dB. We note that the dispersion between the different ratios is lower than $\pm$ 0.5 dB.

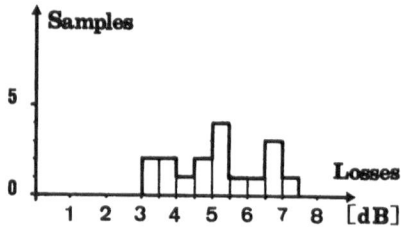

Fig. 3 Histogram of insertion losses in laboratory fabrication

Table 1 : Intensity ratios and insertion loss of a 16-port divider

1	2	3	4	5	6	7	8
16.8	16.1	17.1	16.8	16.5	16.7	16.9	16.5

Intensity ratios

9	10	11	12	13	14	15	16
16.1	16.2	16.8	16.4	16.3	16.8	17	16.4

Intensity ratios

Insertion loss : 4.6 dB

If we excite one of the 16 branches, the ratio is lower than about 0.5 dB in the opposite direction. Figure 3 shows the reproducibility of the method under laboratory conditions. Among 17 independently 8-port fabricated samples, 25 % have total insertion losses between 3 and 4 dB and 50 % between 4 and 6 dB.

4. Conclusion

We have perfected a simple and cheap method for fabrication good quality optical waveguides. The linear losses are 0.7 dB/cm. The insertion losses of a 16-port divider are 4.6 dB. Compared with other similar methods, these results are interesting when considering the simplicity of the process.

References

1 K. Kaede and R. Ishikawa : "A ten-port Graded-index Waveguide Star Coupler Fabricated by Dry Ion Process" ECOC 83 Proc. pp 209-211
2 E. Okuda, H. Wada and T. Yamasaki : "Planar Waveguide 8-port Branched Circuit" 7th Topical Meeting on Integrated and Guided-wave Optics, B6 (1984)
3 F. Cochet : Thèse de Docteur Ingénieur. Grenoble 1982
4 G. Voirin : Thèse de Docteur Ingénieur. Grenoble 1984

Optimization of Ti:LiNbO$_3$ Directional Couplers and Their Application to a Mach-Zehnder Interferometer Modulator

A. Rasch, M. Rottschalk, and W. Karthe

Friedrich-Schiller-Universität Jena, Sektion Physik, Max-Wien-Platz 1
DDR-6900 Jena, German Democratic Republic

The dependence of the coupling strength of Ti:LiNbO$_3$ directional couplers on fabrication parameters has been investigated. An optical intensity modulator was built up with the couplers.

Optical directional couplers produced by the diffusion of titanium strips into lithium niobate crystals are the basis of several guided-wave devices including switches, modulators and wavelength filters. Channel waveguides and couplers have been treated with a variety of theoretical methods [1] - [4]. We report on the dependence of the coupling length of parallel channel waveguides on fabrication parameters. A problem for the optimization was the suppression of the unwanted outdiffusion of Li$_2$O during the diffusion process. These results are basically used to perform a Mach-Zehnder interferometer modulator using a pair of optical gate couplers.

A concise theory of operation of the dual-channel directional coupler is represented by the coupled mode theory of synchronous coherent coupling between the overlapping evanescent tails of the modes guided in each waveguide. A schematic drawing of a directional coupler is shown in Fig. 1. The coupling between modes is given by the general coupled mode equations for the amplitudes of the two modes. Thus,

$$\frac{dA_i(z)}{dz} = -iB_i A_i(z) + c \cdot A_j(z) \qquad i,j = 1,2 \qquad (1)$$

$$i \neq j$$

where B_i are the complex propagation constants of the modes in the two guides and c is the coupling coefficient between modes. We calculated the dependence of the coupling length l on different propagation constants B_i, especially on the optical losses of the waveguides (Fig. 2).

For common values of optical losses in Ti:LiNbO$_3$-waveguides the loss influence can be neglected. But for special applications, the dependence of the coupling length on the waveguide loss may be useful.

Assume that the guides are indentical, the interguide transfer efficiency η is given by the expression

$$\eta = \sin^2 (c L) \qquad (2)$$

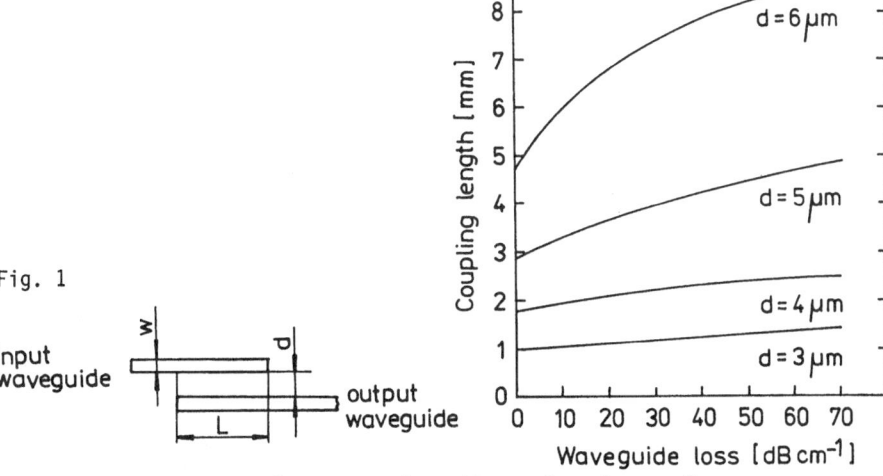

Fig. 2

Fig. 1

input
waveguide

output
waveguide

Fig. 1   Schematic drawing of a directional coupler;
         L is the total interaction length, d is the waveguide
         separation, w is the width of the guide

Fig. 2   Transfer length l vs the waveguide losses for
         different separation distances d

where L is the total interaction length, and c is the coupling coefficient. The interaction length for complete crossover, called the transfer length, is given by $l = \pi/2c$. For
comparison our experimental results with the theory we used
the expression for the transfer length [1]

$$l = l_0 \cdot \exp(d/\gamma) \qquad (3)$$

where l_0 and γ are experimentally determined parameters.
Using the effective index method, the lateral penetration
depth is given by [5] - [6]

$$\gamma = 1/k(2n_s \Delta nb)^{1/2} \qquad (4)$$

where $k = 2\pi/\lambda$, and b is the normalized effective index for
the diffused channel waveguide. So we are able to calculate
the transfer length l in dependence on the effective index,
which is determined experimentally.

   All measurements were made on directional coupler structures fabricated in Y-cut X-propagating lithium niobate. The
waveguide pattern was put onto the crystal by using excellent
electron-beam written masks and mask liftoff technique.
Titanium was deposited by rf-sputtering. The deposited strips
were 3 µm in width and 35 nm in thickness. The metal pattern
were diffused into the crystals at 1273 K for 4 h under
flowing wet argon-/oxygen-atmosphere. We found out that the
Li_2O-outdiffusion can be suppressed by realization of a certain mass current density of the water in the quartz tube

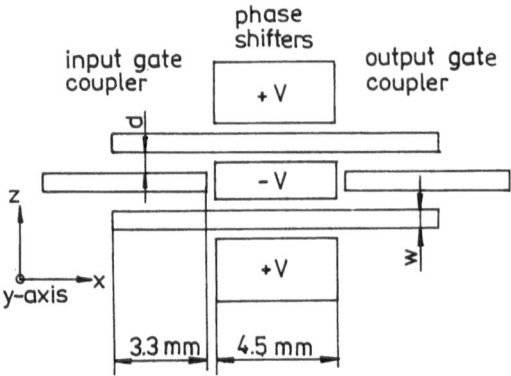

Fig. 3   Top view of the Mach-Zehnder interferometer modulator
         w is the strip width: 3 /um, d is the separation
         distance: 4 /um

where diffusion takes place. The mass current density J can
be expressed as

$$J = f \cdot I_V \cdot A^{-1} \qquad (5)$$

where f is the absolute humidity of a gas, I_V is the gas
flow and A is the cross-section of the tube. Outdiffusion can
be suppressed by a value $J \geqslant 1.7 \times 10^{-2}$ g $(cm^2 \cdot min)^{-1}$ for
diffusion times $T_D \leqslant 10$ h (the result is in agreement with
values reproduced from the literature).

For each measurements of the transfer length l, a series
of twenty couplers of increasing length were fabricated on
the same crystal. The measurements were made using TE polarized light, which sees the extraordinary refractive index.
The ends were polished very well to allow endfire coupling.
For purposes of determining l each waveguide pair was assumed to be phase-matched, which appeared valid since transfer
efficiencies approaching 100 % were frequently observed. So
we measured the transfer length l as a function of the
separation distance d and the effective waveguide index.

All these results were used to build up a Mach-Zehnder
interferometer modulator comprising a pair of optical gate
couplers and conventional phase shifters. The optical gate
coupler consists of three coupled straight waveguides
(Fig. 3). The light intensity in the input waveguide is
equally divided (1 + 0.005 : 1 + 0.005) at an input gate
coupler and coupled into parallel arms of the phase shifters.
The retardation between the two arms of the phase shifters
is controlled electro-optically by applied voltage. At the
output gate coupler, both light re-combined according to the
phase shifters. The following experimental results were obtained:

Modulation depth:	95 %
Half voltage:	1.7 V
Maximum of frequency:	1 GHz
Wavelength:	0.6328 /um
Optical throughput loss:	3 dB

References:

1   E.A. J. Marcatilli: Bell. Syst. Tech. J. 48, 2071 (1969)
2   R.C. Alferness, R.V. Schmidt, E.H. Turner: Appl. Opt.
    18, 4012 (1979)
3   M. D. Feit, J. A. Fleck, L. McCaughan: J. Opt. Soc. Am.
    73, 1296 (1983)
4   J. Ctyroky, M. Hofman, J. Janta, J. Schröfel: IEEE J.
    Quantum Electron. QE-20, 400 (1984)
5   G. B. Hocker, W. K. Burns: IEEE J. Quantum Electron.
    QE-11, 270 (1975)
6   G. B. Hocker, W. K. Burns: Appl. Opt. 16, 113 (1977)

Measurement of Guided Mode Cut Off Wavelengths in Ti:LiNbO₃ Channel Waveguides

K. Thyagarajan*, A. Enard, P. Kayoun, D. Papillon, and M. Papuchon

THOMSON-CSF Laboratoire Central de Recherches, Domaine de Corbeville, B.P. 10 F-91401 Orsay Cedex, France

Measurements on the variation of the cut off wavelengths of the (0,0) and (0,1) modes with Ti strip width in Ti:LiNbO₃ channel waveguides are presented. Comparison with theoretical estimation using effective index method is shown to be reasonably good. Such measurements must find applications in integrated optical device optimization.

Optical channel waveguides form the basic element in most integrated optical devices. Most of these devices like the Mach-Zehnder interferometer modulator, directional coupler switch, X-switch etc. are based on either single mode or two mode waveguides [1-8]. In order to optimize their performances at a given wavelength of operation, it is important to know the region of single mode operation of these waveguides. We report here measurements of the cut off wavelength λ_c of (0,0) and (0,1) modes in Ti:LiNbO₃ channel waveguides using the spectral transmission method used often in characterizing single mode fibers [9]. Variation of λ_c with Ti strip width before diffusion is presented. A theoretical estimation using the effective index method [10] is shown to give a reasonably good agreement. Such cut off wavelength measurements must be useful in integrated optical device optimization and as a feedback in device fabrication.

The experimental set up used is shown in Fig. 1. Light from a tungsten halogen lamp is focussed on the entrance aperture of a monochromator after passing through a chopper. The exit slit of the monochromator is imaged on the input end of the channel waveguide. The output of the waveguide is imaged on a liquid nitrogen cooled

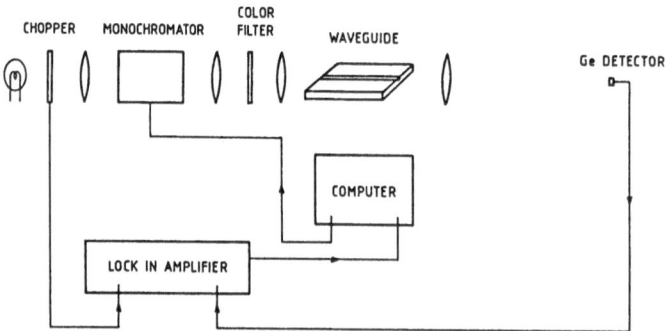

Fig. 1 The experimental set up used for measuring the cut off wavelength.

---

* On leave from Department of Physics, Indian Institute of Technology, NEW DELHI 110016 INDIA

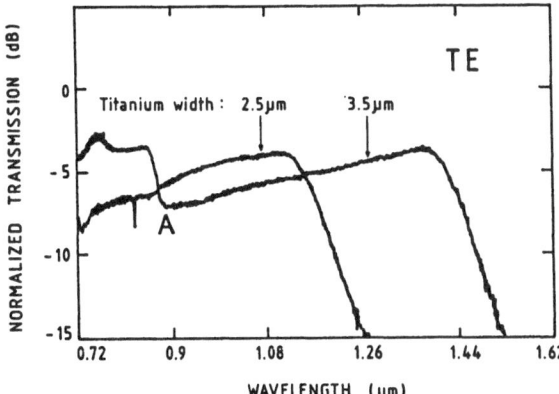

Fig. 2 Typical normalized TE mode transmission spectra of channel waveguides
formed by diffusing 2.5 µm and 3.5 µm Ti strip in Z cut LiNbO3.

germanium detector whose output is connected to a lock in amplifier. The monochromator is driven by a computer which simultaneously collects data on the output
intensity. This output spectrum is normalized with respect to the source spectrum
which is measured without the waveguide. Typical normalized transmission spectra
for two different Ti width channel waveguides are shown in Fig. 2. The similarity
with the transmission spectrum of a single clad optical fiber is apparent. The only
difference is in the presence of a cut off even for the fundamental mode for such
asymmetric waveguides.

We define the cut off of the (0,1) mode to be at the point where the normalized
transmission spectrum starts to rise again (point A in Fig. 2). For the cut off of the
(0,0) mode we take arbitrarily the wavelength at which the transmission drops by 3 dB
after attaining a maximum.

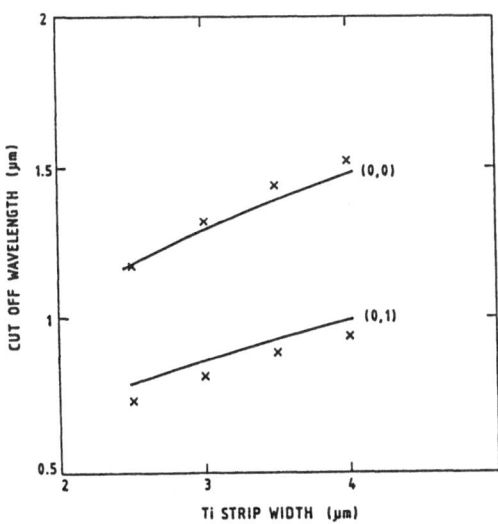

Fig. 3 Variation of the cut off wavelength of the (0,0) and (0,1) modes as a function
of the Ti strip width before diffusion. The crosses represent the experimental
points and the curve corresponds to effective index calculations.

Figure 3 shows the variation of λ_c of the (0,0) and (0,1) modes with the Ti strip width before diffusion for the TE modes in Z-cut Ti:LiNbO$_3$ waveguides fabricated with Ti thickness $\sim$ 600 Å, diffusion temperature $\sim$ 1050°C, diffusion time $\sim$ 8 hours.

For a theoretical estimation we have used the effective index method [10] to calculate the cut off wavelengths. In order to estimate the values of (dn/dc) and the diffusion constant D_o we fabricated a planar waveguide in Z cut LiNbO$_3$ and measured the effective indices of the various modes using the prism coupling technique. Then a fitting to a Gaussian index profile gave us the following values :

$$(dn/dc)_o \sim 1.02 \times 10^{-29} \text{ m}^{-3}$$
$$(dn/dc)_e \sim 1.75 \times 10^{-29} \text{ m}^{-3}$$
$$D_o \sim 1.7 \times 10^{-6} \text{ m}^2/\text{s}$$

These values are consistent with those of other authors [11,12,13].

For fitting the theoretical curve to the experimental points, the number of Ti atoms per unit area available for diffusion (i.e., atomic density of Ti in the film multiplied by the Ti film thickness) was varied in order to fit at one point. Then the curves were calculated by simply varying the Ti strip width. If we assume the atomic density of Ti in the film to be the same as that in solid Ti (i.e, 5.64×10^{28} m^{-3}) then the theoretical fit in Fig. 3 corresponds to a Ti film thickness of 630 Å. The wavelength dependence of the substrate refractive index was estimated using the Sellemier fit given by Rauber [14].

From Fig. 3 it can be seen that the agreement is fairly good in spite of the fact that the effective index method is not expected to be very accurate near cut off.

In conclusion, we have extended the method for measurement of the cut off wavelength from spectral transmission measurement to the case of diffused channel waveguides in integrated optics. Theoretical estimation using the effective index method is shown to give good agreement. The method should find application in device optimization and as a feedback in integrated optics.

Acknowledgements

Thanks are due to J.M. Arnoux, M. Werner and R. Bourmaleau for their expert technical assistance. This work was partialy supported by DRET.

REFERENCES

1   R.C. Alferness, IEEE J. Quantum Electron. QE-17, (1981) 946.

2   M. Papuchon, Y. Combemale, X. Mathieu, D.B. Ostrowsky, L. Reiber, A.M. Roy, B. Sejourne, and M. Werner, Appl. Phys. Letts. 27, (1975) 289.

3   R.V. Schmidt and H. Kogelnik, Appl. Phys. Letts. 28, (1976) 503.

4   R.C. Alferness, IEEE Trans. Micr. Th. Tech., MTT-30, (1982) 1121.

5   M. Papuchon, A.M. Roy and D.B. Ostrowsky, Appl. Phys. Letts., 31, (1977) 266.

6   A. Neyer, Electron. Letts, 19, (1983) 553.

7   A. Neyer, IEEE, J. Quantum Electron., QE-20, (1984) 999.

8 M.D. Fiet, J.A. Fleck and L. Mc Caughan, J. Opt. Soc. Am., 73, (1983) 1296.

9 K.A.H. Van Leeuwen and H.T. Nijnuis, Optics Letts. 9, (1984) 252.

10 G.B. Hocker and W.K. Burns, Appl. Opt. 16, (1977) 113.

11 C.H. Bulmer and W.K. Burns, J. Lightware Tech, LT-1, (1983) 227.

12 S.K. Korotky, W.J. Minford, L.L. Buhl, M.D. Divino, R.C. Alferness, IEEE J. Quant. Electron., QE-18, (1982) 1796.

13 M. Fukuma and J. Noda, Appl. Opt., 19, (1980) 591.

14 A. Rauber, Current Topics in Material Science Vol I, Kaldis, E (Ed.), North Holland, (1978) 481.

Index of Contributors

Springer Series in Optical Sciences

Editorial Board:

J. M. Enoch,
D. L. Mac Adam,
A. L. Schawlow,
K. Shimoda,
T. Tamir

Managing Editor:
H. K. V. Lotsch

Springer-Verlag
Berlin
Heidelberg
New York
Tokyo

B. R. Frieden

Probability, Statistical Optics, and Data Testing

A Problem Solving Approach

1983. 99 figures. XVII, 404 pages. (Springer Series in Information Sciences, Volume 10)
ISBN 3-540-11769-5

Contents: Introduction. – The Axiomatic Approach. – Continuous Random Variables. – Fourier Methods in Probability. – Functions of Random Variables. Bernoulli Trials and its Limiting Cases. – The Monte Carlo Calculation. – Stochastic Processes. – Introduction to Statistical Methods: Estimating the Mean, Median, Variance, S/N, and Simple Probability. – Estimating a Probability Law. – The Chi-Square Test of Significance. – The Student t-Test on the Mean. – The F-Test on Variance. – Least-Squares Curve Fitting – Regression Analysis. – Principal Components Analysis. – The Controversy Between Bayesians and Classicists. – References. – Subject Index.

Integrated Optics

Editor: **T. Tamir**

With contributions by E. Garmire, J. M. Hammer, H. Kogelnik, T. Tamir, F. Zernike

2nd corrected printing of the 2nd corrected and updated edition. 1982. 99 figures. XV, 339 pages. (Topics in Applied Physics, Volume 7).
ISBN 3-540-09673-6

Semiconductor Devices

for Optical Communication

Editor: **H. Kressel**

2nd updated edition. 1982. 191 figures. XIV, 309 pages. (Topics in Applied Physics, Volume 39). ISBN 3-540-11348-7

Springer-Verlag
Berlin
Heidelberg
New York
Tokyo

Springer Series in Optical Sciences

Editorial Board: J.M. Enoch D.L. MacAdam A.L. Schawlow K. Shimoda T. Tamir